Grasses:
a guide to identification using vegetative characters

By Hilary Wallace

Ecological Surveys (Bangor)

© Text Field Studies Council 2021, minor text revisions 2023;
all photographs Hilary Wallace 2021, except where credited;
line art Rebecca Farley-Brown 2021
ISBN 978 1 908819 55 0
OP189

Contents

Dedication	iv
Acknowledgements	iv
Introduction	1
Getting started	4
A key to common grass species	12
Glossary	97
Further reading	102
Index of species accounts	103

Dedication

To Mike Prosser (1934-2016). Mentor, friend and inspiration.

Acknowledgements

The idea of developing this AIDGAP vegetative key to grasses was first proposed by Sue Townsend, and its development has been helped by innumerable students over many decades of tutoring.

More specifically in finding sites to photograph the less common species or providing specimens: Mag Cousins, Caroline O'Rouke, Mark Duffell, Simon Smart; and testing that the key works: Owen Mountford, Mag Cousins.

Line illustrations by Rebecca Farley-Brown.

Photographs by Hilary Wallace unless indicated.

Introduction

Background to the key

This guide is aimed at habitat surveyors who need an easily accessible key to common grass species and also to those less common species that are diagnostic of more restricted habitats.

The key started life as a simplified key to agricultural grasses, developed at the Agricultural Botany Department at Bangor University. The key was subsequently expanded and developed by Mike Prosser in the 1970s to cover 62, principally grassland, species.

Over the past 30 years the key has been expanded and used extensively for teaching students on habitat survey courses; at the Snowdonia National Park Study Centre, the Field Studies Council and courses commissioned by Government agencies requiring staff to conduct habitat surveys and assessments. Courses have covered the National Vegetation Classification (NVC, Rodwell 1991 et seq.), Phase I Habitat survey (JNCC 1990), farm stewardship surveys and Habitat Condition Assessments. Often students attending these courses had limited experience of observing or identifying grasses, claiming they are 'too difficult' and thus largely ignored, despite being essential components of most habitat survey protocols. Feedback from students has resulted in many of the more difficult couplets in the key being revised, new features have been added and a number of species can be keyed out via more than one route. Most recently the key has been trialed as a stand alone course on *Vegetative Grass Identification* at the Field Studies Council.

The popularity of the key is evident by the number of requests for copies from past students and their colleagues, and the feedback from surveyors who are still using it many years after completing their course.

Although it now covers 90 species the key retains the structure of a single dichotomous key, rather than the use of sections, now so common in keys with ever expanding numbers of species.

The structure and order of species remains largely the same as previous versions so surveyors who are familiar with earlier versions should have no problems adjusting to the expanded number of species covered here. The main changes are the addition of saltmarsh and sand dune species, and also species of more restricted habitats (e.g. *Sesleria caerulea* in calcareous grasslands) or geographical ranges (e.g. *Agrostis curtisii* in heathlands in SW England and S Wales). Still loosely based on the dichotomous key of Hubbard species not covered in this guide should be relatively easy to track down using the Hubbard (1984) key.

The dichotomous key is accompanied by line illustrations of the diagnostic features referred to in the key, coloured photos of most species and notes on distinguishing features, habitat preferences and current distribution and status in UK, the latter gleaned mainly from the *Atlas of the British Flora* (Preston *et al.* 2002).

Why a vegetative key?

Flowering grasses are difficult to identify as the flowers are very small and the keys usually require microscopic examination of the individual floral parts. When in flower the structure and arrangement of the floral parts into different panicle formations provides an initial guide to the species group, and sometimes to the species, but usually there is no substitute for careful microscopic examination of the floral parts. With more and more alien species being included in many of the recent keys they are becoming increasingly complex for the beginner. *The Flora of the British Isles* published in 1987 (Clapham *et al.* 1987) covered just 58 genera of grasses, whilst the accompanying excursion flora (Clapham *et al.* 1995) covered the 41 most common ones. In the second edition of Stace (Stace 2005) the number of genera had increased to 94, rising to 100 in the third edition (Stace 2010). With this increase in genera both the latest BSBI guide (Cope and Grey 2009) and Stace (2019) have in excess of 220 species, and the keys are divided into a number of sections based on tribes; unfamiliar territory to most field surveyors.

In addition to the problems of using the floral keys for identification much habitat assessment work carried out by surveyors will involve grasses without flowers. There are a number of reasons for this:

- habitat surveys are carried out all year round
- species have different flowering periods, even within the same habitat;
- habitat management will affect flowering (e.g. cutting, grazing, etc.) such that flowers have often been removed at the time of survey,
- unfavourable conditions e.g. dense shade or competition from more aggressive species, will inhibit flowering.

All these factors potentially restrict survey accuracy if reliance is placed on identification using flowering material.

Until recently Hubbard (1984) provided the only comprehensive key based entirely on vegetative characteristics. Although Hubbard's key works very well the layout has proved daunting for most beginners. For the less common species that are not covered in this publication reference to Hubbard will usually enable rapid identification as the basic structure and order of species in the two keys is very similar, even if the layout is different.

There have been few attempts since Hubbard to produce a comprehensive vegetative key. Rose (1989) has a key to a limited number of species, divided by habitat, but it is far from comprehensive. Poland and Clements (2020) provide a very comprehensive key to over 200 species; including aliens, introductions, crop species, hybrids and rarities. The number of species and the terminology used by Poland can be challenging to the beginner, and in most habitat survey work the majority of grasses encountered are common. Poland's keys will prove useful to the more advanced surveyors and those working in disturbed or unusual habitats; and they provide additional characters that can help in difficult situations.

Fitzpatrick *et al.* (2016) have produced a key to Ireland's grasses that includes a vegetative key to 75 species based on the keys of Hubbard (1984) and Poland and Clements (2009). Their key groups species into 6 sections based on simple characteristics. There are probably many other local keys produced as teaching aids throughout the botanical network.

Why are grasses important?

Grasses form the matrix of a wide range of habitats in Britain; from species-rich calcareous grasslands to species-poor saltmarshes and mobile sand dune systems. They also provide important cover in urban landscapes including road verges, parkland and playing fields. In some habitats grasses play an active role in habitat creation through production of extensive rhizome systems, for instance in stabilizing mobile sand in dune systems and mud flats. In other situations, if left unmanaged, dense tussock forming species can threaten the conservation value of species-rich habitats through unchecked growth. Thus, the nature of growth and colonisation potential of the different species can play a critical role in habitat development and conservation. Some species have a wide ecological tolerance whilst others are very restricted, thus species can be an important aid in habitat classification.

The accurate classification of many habitat types, not only grasslands, using the Phase I habitat scheme depends on the ability to recognise a relatively small number of grass species whilst for a more detailed classification using the National Vegetation Classification scheme a greater number of species are required. Habitat quality will also be linked to the species present and their relative abundance, and although some species are very common with a wide tolerance to soil chemistry and moisture status others are restricted to narrow conditions and for many of these their presence is of conservation importance.

Limitations

The key does not aim to be comprehensive, but covers most grass species likely to be encountered by the average surveyor carrying out habitat surveys.

Recent introductions, aliens, crop species and hybrids are not included.

Getting started

What you need

- A good hand lens is essential for observing the structure and features referred to in the key; x 10 or ideally x 20 for observation of hairs, etc.

- Have a ruler to hand for measuring leaf width, etc.

- Always collect 'good material'. Don't rely on a single shoot or half a leaf! Always make sure you observe growth form and if necessary collect 'the whole plant', including complete leaves and shoots together with root, runners, etc.

- The features used in the key refer to the vegetative shoot only, beware not to use the flowering culms as characters on these are often different; especially in relation to ligules.

What characters to look for and in what order

It is important to be able to list the key features of a specimen before attempting the key.

When using this key it is essential to restrict your observations to the vegetative shoots (tiller).

- A general character of grasses that separates them from the sedges is the usually hollow, cylindrical stem and the arrangement of leaves being alternate. Leaf blades are usually linear, entire and thin and on vegetative shoots the leaf blades have long sheaths covering the stem.

- Vegetative identification relies on detailed observation of a relatively small number of easily observed characters.

- A series of figures give a schematic illustration of the critical features to observe on a grass before embarking on using the key.

Anatomy of the vegetative grass

Figure 1 illustrates the anatomy of a grass plant and distinction between flowering culms and vegetative shoots.

For this key restrict your observations to the vegetative shoots (tiller). Some books refer to these vegetative shoots as innovations (Cope and Grey 2009, Stace 2010). The order of observation for using the key is:

1. **Leaf form**: leaf blade width (narrow like a needle or a wider blade), leaf blade structure (rolled or folded)

2. **Ligule**: Present or absent

3. **Auricle**: Present or absent

4. **Leaf sheath**: open or closed, degree of hairiness or not hairy

5. **Life cycle**: annual or perennial

6. **Growth form**: propagation – are the shoots dispersed or clump forming.

The variety of forms for these characteristics are illustrated by a series of diagrams and photographs.

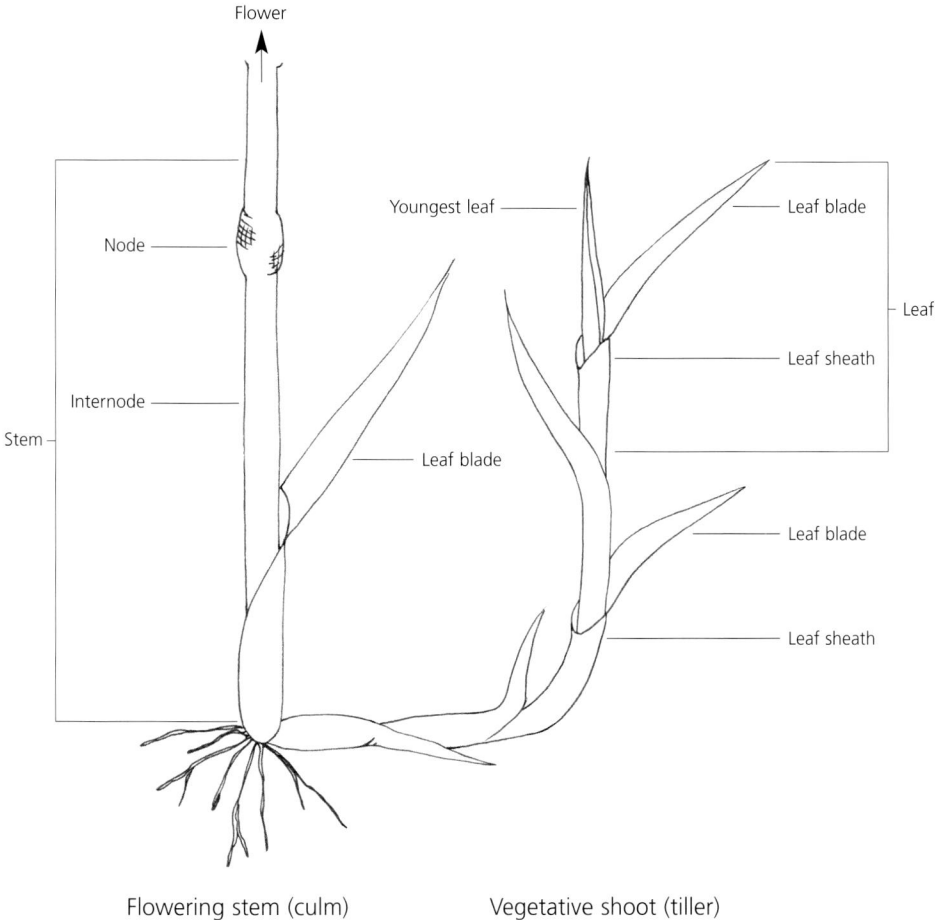

Figure 1. Distinction between flower stems (culms) and vegetative shoots (tillers).

1. Leaf form

The leaf is comprised of a blade and a sheath (Fig. 1).

- The leaf blade is that part above the sheath, sometimes referred to as the lamina, often flat but sometimes bristle-like.
- The leaf sheath is the extended lower part of the leaf surrounding the stem.

> **The structure of the leaf blade forms the first dichotomy in the key.**
>
> • Does the leaf open out into a flat blade? If a flat blade is it rolled or folded in the shoot? (Fig. 2a-c.)
>
> • or is the leaf blade very narrow like a needle, this usually means it is tightly inrolled or grooved? (Fig. 2d.)

Figure 2. Leaf blade flat versus bristle-like.
a. leaf usually flat (opened out); b. leaf folded; c. leaf rolled;
d. leaf blades bristle-like.

2. Ligule

The second question in the key refers to the ligule.

A ligule is a membranous outgrowth that arises on the inside of the leaf blade where it joins the leaf sheath (Figs 3, 4).

- Are ligules present or not?
- If there is a ligule, what form does it take?

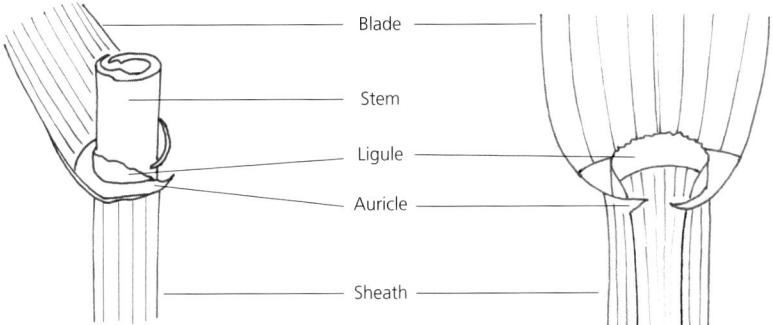

Figure 3. Ligule and auricle, at junction of leaf blade and leaf sheath.

- Is it reduced to a ring of hairs rather than a membranous outgrowth? (Fig. 4a)
- If membranous then the main questions are:
    - is the ligule broader than it is long?
    - is the tip blunt or pointed?
    - is the tip smooth or ragged?

Some examples are given below, more will be illustrated as you progress through the key.

- Longer than broad and pointed (Fig. 4b)
- Shorter than broad and blunt (entire) (Fig. 4c)
- Longer than broad with a ragged top (Fig. 4d)

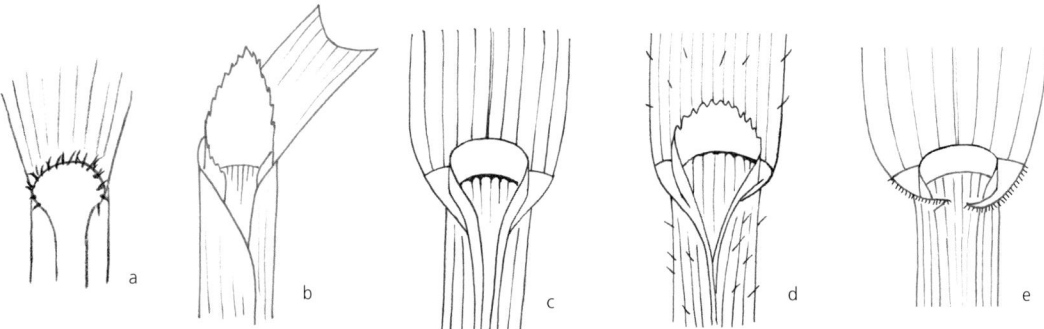

Figure 4. Examples of ligules. a. ligule replaced by a ring of hairs. b. longer than broad and pointed; c. shorter than broad and blunt (entire); d. longer than broad with a ragged top; e. auricles with hairs.

3. Auricles

Does the plant have auricles or not? Auricles are claw-like outgrowths at the point where the leaf blade joins the leaf sheath (Fig. 3).

They take a variety of forms and sizes.

- Most commonly they clasp around the leaf sheath (Fig. 3)
- Sometimes they spread out from the sheath
- Occasionally they have hairs on them (Fig. 4e).
- Sometimes they are obscure and you need to observe more than one plant or shoot to be sure of your answer.

4. Sheath

The leaf sheath is the lower part of the leaf surrounding the stem. Sheaths can be **open** or **closed**.

- In some species the leaf margins are wrapped around the stem, the edges of the leaf being clearly visible as they run down the stem, referred to as an **open** sheath (Fig. 5a).
- In others the margins of the leaf are fused to form a tube around the stem, referred to as a **closed** sheath (Fig. 5b).

Imagine the difference between a woolly cardigan that wraps around (open) compared to a v-necked jersey (closed); or a scarf compared to a neckwear tube.

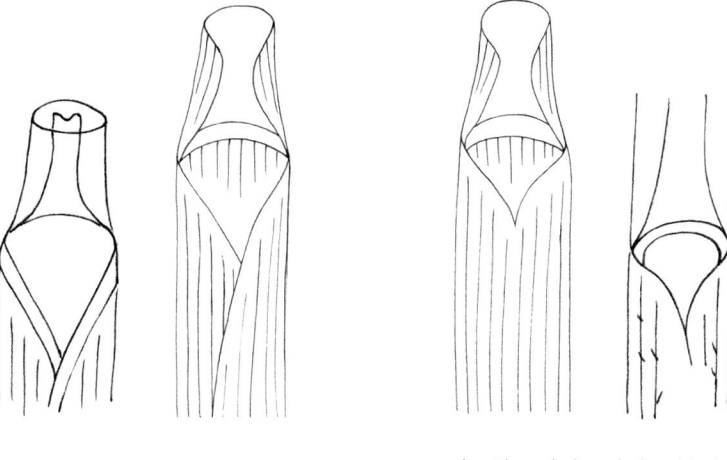

a. Open sheath
free margins

b. Closed sheath (and hairy)
fused margins

Figure 5. Example of open versus closed leaf sheaths.

- Is the sheath hairy? These hairs may be sparse or dense; closely appressed to the stem or spreading, restricted to just the lower part of sheath, etc.

5. Life cycle

Annual or perennial? **Perennial** species tend to accumulate dead material around their bases whilst annuals do not. **Annuals** are often only encountered in flower due to their short life cycle whilst many perennials will be found throughout the year with only vegetative shoots, especially in habitats that are regularly managed through grazing or mowing.

6. Growth form

If perennial does it form tight clumps or are the tillers spread out through a network of extensive rhizomes or stolons.

- Plants may form tight, compact clumps (**tussocks**), which usually indicates that they do not have extensive rhizomes or stolons. Good examples of dense tussock forming species are *Dactylis glomerata*, *Deschampsia cespitosa* and *Molinia caerulea*.

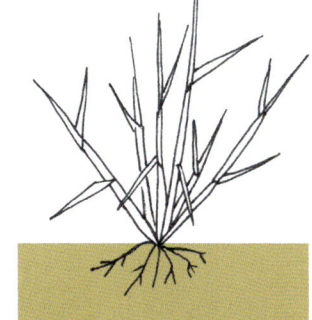

- **Stolons** are stems that creep along above the ground surface. They often produce roots at nodes along these stems and from these new vertical shoots arise, *Agrostis stolonifera*, *Alopecurus geniculatus*.

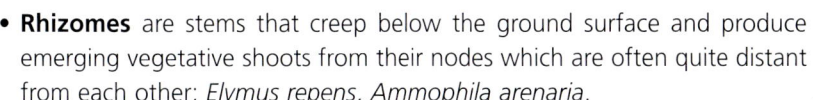

- **Rhizomes** are stems that creep below the ground surface and produce emerging vegetative shoots from their nodes which are often quite distant from each other; *Elymus repens*, *Ammophila arenaria*.

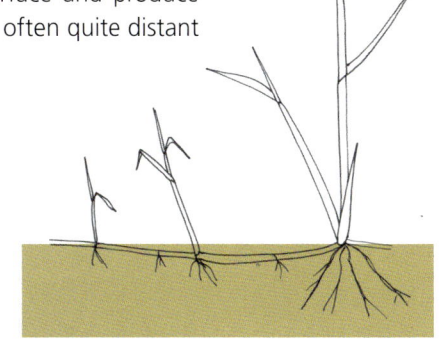

How to use the key

- The means of identifying an unknown specimen is through a traditional dichotomous key. The key presented here has not been sub-divided into sections (as have so many modern keys), so if you go wrong it is relatively easy to track back to the question that you were uncertain of.

- The numbers on the left hand side are referred to as **couplets**, they provide an either/or option for you to answer.

- You need to start at Number 1 and decide which description best fits your specimen. Read the alternatives carefully as sometimes there are a number of characters that when combined give the best description. Once you have decided which is closest to your sample read across to the number on the right hand (RH) side of the key that goes with your choice of answer and move on to that couplet number. For example, if you start with a leaf blade that is **flat** you would go from couplet 1 to couplet 2 whilst if your leaf was narrow and bristle-like you would go to couplet 99.

- Couplet 2 then asks whether you have a ligule that is a fringe of hairs, in the absence of auricles – that takes you on to Couplet 3. If you have a membranous ligule, or no ligule, you would proceed to Couplet 8 (see Table 1).

- The main dichotomies have been highlighted in bold to make tracking your way through the key easier.

- You carry on progressing through the key in this way until you reach a description that matches a species name on the right-hand side.

- Once you reach a name additional information is provided about diagnostic features compared with other similar species, habitat preferences, distribution, etc.

- There have been a lot of changes in nomenclature for grasses so in addition to a common name and the currently accepted latin name (using Stace 2019), other names that have been regularly used, with appropriate references, are also given. This is particularly relevant if you are carrying out NVC surveys where many of the names used in the British Plant Communities volumes have changed, sometimes more than once, since its publication in the 1990s (Rodwell 1991 *et seq*.). Previously commonly used names are indicated by: • Hubbard 1984; • Clapham, Tutin and Moore 1987; • Stace 2010.

An illustrated glossary is provided at the end of the key.

It is worth becoming familiar with the main divisions of the key, these couplets have been indicated by bold type in the key and with increasing familiarity will allow a fast track through the key.

A brief synopsis of the main divisions is given below.

Table 1.

Couplet	Character	Goes to Couplet
1	Leaf blade flat	2
1	Leaves needle-like	99
2	Ligule a ring of hairs	3
2	Ligule membranous	8
8	Leaf blade rolled in shoot	9
8	Leaf blade folded	78
9	Leaves with auricles	10
9	Leaves without auricles	23
23	Leaf sheath open, with free margin	24
23	Leaf sheath tubular	67/68
68	Leaf sheath hairless	69
68	Leaf sheath hairy	72
78	Lower sheath hairy	79
78	Lower sheath hairless	81
99	Annual	100
99	Perennial	104

Colour photos of species included in the key, and additional pictures illustrating other diagnostic and habitat characteristics, are also available on line at:

www.fscbiodiversity.uk/id-resources-built-fsc-identikit

where they are linked to a multi-access key.

A key to common grass species

Couplet	Figures

1a. Leaf blades usually flat 2

1b. Leaf blades bristle like, narrow 99

Leaf blades usually flat ▼

2a. Ligule formed by a dense fringe of short hairs; no auricles present at the junction of sheath and blade 3

2b. Ligule membranous or absent 8

Ligule formed by a dense fringe of short hairs ▼

3a. Leaf blade 10-30 mm wide, often glaucous; rhizomes long; plant reed-like. Found in wet places or in water ....
.... ***Phragmites australis***
Common Reed

3b. Lead-blade <15 mm wide; with or without rhizomes; plants not reed like 4

Identification notes / key diagnostics

Phragmites australis

Large reed-like perennial up to 3 m tall. Leaves up to 60 cm long and 30 mm wide. Ligule hairs short but often with long (up to 8 mm) lateral whiskers. Isolated shoots connected by long underground rhizomes, but may form extensive and dense stands.

H Swamps, fens, open water, ditches and may expand into wet grassland. Also on coastal cliffs.

D Thoughout. Mainly lowland but up to 450 m on Clee Hill, Shropshire.

S Common and stable. Expanding due to eutrophication and often planted for wildlife benefits.

4a. Youngest leaf usually folded in the sheath. Leaf sheath bearded with short spreading hairs at the junction of the sheath and blade; leaves narrow, blades stiff with blunt tips. Heaths and moors .... **Danthonia decumbens**
(*Sieglinglia decumbens* •)
Heath Grass

4b. Youngest leaf rolled in sheath. Leaves hairless on the outside at the junction of leaf and blade, dull green 5

5a. Leaves 3-10 mm wide, slightly hairy or hairless, finely pointed; tufted perennial, often forming raised hummocks; roots cord-like. Damp, often peaty habitats .... **Molinia caerulea**
Purple Moor-grass

5b. Blades smooth, hairless; shoots cylindrical; plants with creeping rhizomes. Saltmarshes 6

6a. Leaf blade narrow <6 mm; ligule hairs short, 0.2-0.6 mm; rhizomes short, wiry. Saltmarshes .... **Spartina maritima**
Small Cord-grass

6b. Leaf blade wide, 6-15 mm; ligule hairs longer, up to 3 mm; rhizomes long, stout. Coastal mud flats 7

Danthonia decumbens

Tuifted perennial. Leaves up to 25 cm long and 2 mm wide, often folded when young and blunt tipped, sometimes sparsely hairy. Leaves are generally stiff and often curved downwards, often slightly glaucous. In additon to the hairs in place of the ligule it also has spreading hairs at the junction of the leaf blade and sheath.

🄷 Sandy and peaty soils on moorland, heaths and rough grassland, occasionally in unimproved pastures, generally on moist soils.

🄳 Thoughout; lowland to montane.

🅂 Native but declining in the east through habitat loss.

Molinia caerulea

Densely tufted perennial forming tight clumps which often develop into raised tussocks. cf *Danthonia*: leaves of *Molinia* are longer (up to 45 cm) and wider (3-19 mm) always rolled in the youngest shoots; taper to long, narrow points and are not stiff.

🄷 Wet and damp peaty soils or heavy clays; on moorland, heaths, fens, and damp birch woodlands; usually where there is some free water movement in the soil profile. Rapidly colonises after burning or following abandonment. Marshy grassland indicator.

🄳 Throughout, but relatively scarce in the east Midlands and SE England.

🅂 Declining in many lowland areas but expanding in areas of reduced grazing in the uplands.

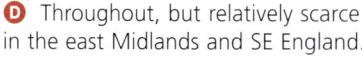

Spartina maritima

Rhizomes are short so tufts and patches tend to be smaller in extent than those of the other two species of *Spartina*. Ligule hairs are also shorter. Blades only slightly spreading.

🄷 Tidal muds and sands in bare areas by the sea and in estuaries. Generally higher up the shore than *S. anglica*. It occurs fringing gullies, in shallow pans, and sometimes behind the sea wall.

🄳 Local in south and east England, north to Lincolnshire. Introduced in Ireland and Dorset.

🅂 Native and declining. Losses due to habitat loss and also the expansion of *S. anglica*.

Miguel Sanz Alcántara – Own work, CC BY 3.0, https://commons.wikimedia.org/w/index.php?curid=7784349

7a. Ligule hairs long (2-3 mm); leaf blades widely spreading .... ***Spartina anglica***
Common Cord-grass

7b. Ligule hairs shorter (1-2 mm); leaf blade only slightly spreading .... ***Spartina* x *townsendii***
Townsend's Cord-grass

Ligule membranous or absent (from couplet 2) ▼

8a. Youngest leaf blade rolled in the shoot 9

8b. Youngest leaf blade folded length-wise about the middle nerve in the shoot 78

Youngest leaf blade rolled in the shoot (from couplet 8) ▼

9a. Leaves with small, narrow, claw-like outgrowths (auricles) at junction of sheath and blade 10

9b. Leaves without such auricles 23

Spartina anglica

Deep rooting with very stout, long fleshy rhizomes forming large clumps and meadows. Ligule hairs silky ciliate and longer than the other two species of *Spartina*. Leaf blades spread widely from the main stem.

H Coastal mud flats and saltmarshes.

D Coasts and estuaries around England, Wales and Ireland. Scarce in south Scotland, absent from the north.

S A native derivative of *S* x *townsendii* first recorded in Southampton Waters c.1890. Spread naturally and through planting as a mud stabiliser.

Spartina x townsendii

Similar to *S. anglica* but rhizome scaly and ligule hairs shorter. Leaf blades tend to be more erect, only spreading slightly from the main stem.

H Tidal flats and saltmarshes.

D Coasts and estuaries, much planted to stabilise mud.

S Mainly southern England with scattered locations further north.

Jürgen Howaldt / CC BY-SA 2.0 DE
(https://creativecommons.org/licenses/by-sa/2.0/de/deed.en)

10a. Plants with extensive rhizomes; leaves often grey-green ... 11

10b. Plants without rhizomes, loosely or densely tufted; annual or perennial 13

11a. Leaf blade not ribbed on upper surface, nerves widely spaced. Wide habitat range but rarely on sand .... ***Elymus repens***
(*Agropyron repens* • *Elytrigia repens* •)
Couch Grass

11b. Leaf blade ribbed on upper surface, nerves closely spaced, <1/2 their own width apart, greyish or bluish grey; sheaths with short bristles on their edges. Restricted to sand dunes and saltmarshes 12

12a. Leaf blade wide, 8-20 mm, usually flat; auricles pronounced, clasping/overlapping; rhizomes stout. Sand dunes .... ***Leymus arenarius***
(*Elymus arenarius* •)
Lyme Grass

12b. Leaves narrow, 2-6mm, usually inrolled; auricles short and narrow; rhizomes wiry. Sand dunes, saltmarshes, and banks of brackish water courses .... ***Elymus athericus***
(*Agropyron pungens* • *Elymus pycnanthus* • *Elytrigia athericus* •)
Sea Couch

Elymus repens

Nerves in the leaf are ill defined and widely spaced. Auricles are well defined. Leaves often slightly hairy. Rhizomes can be very extensive and quite deep.

🄷 Fertile, often disturbed habitats, including grasslands, roadsides, waste places, arable fields and coastal areas around seawalls, sand dunes and margins of saltmarshes.

🄓 Throughout. Less common in Highland Scotland.

🅂 Native and stable.

Leymus arenarius

A striking grass. Characterised by wide, blue-grey, stout leaves forming large patches or clumps. Main difference from *E. athericus* is its wider, usually flat, leaf blade, up to 20 mm wide and well defined, long auricles. Rhizomes are much stouter than in *E. repens*. Re *Ammophila arenaria* which occurs in similar habits it has only a short ligule compared to that of *Ammophila* which can reach 3 cm.

🄷 Sand dunes, especially in the mobile foredunes where the deep rhizomes help bind the sand together.

🄓 Throughout coastal regions.

🅂 Native and stable.

Elymus athericus

Blades narrower than *Leymus arenarius*, only to 6 mm wide, and sharply inrolled. Auricles shorter and spreading compared to *L. arenarius*.

🄷 Brackish habitats; creeks, saltmarshes, sand dune-saltmarsh transitional zones, and shingle banks.

🄓 England and Wales; southern Scotland. Absent from north Scotland.

🅂 Native and stable.

13a. Lower leaf sheath hairy; upper sheath hairy or not 14

13b. Lower leaf sheath not hairy; upper sheath hairy or not 18

14a. Leaf sheath stiffly hairy, hairs up to 4 mm; sheath tubular but soon splitting open; loosely tufted perennials 15

14b. Lower leaf sheath softly or shortly hairy, often short – obscure (<0.5 mm); upper sheath not hairy .:.................... 16

15a. Ligule long (to 6 mm) jagged; auricles small; sheath hairy throughout; blade dull green, sparsely hairy or hairless. Woods and shady places .... ***Bromopsis ramosa***
(*Bromus ramosus* • *Zerna ramosa* •)
Hairy-brome

15b. Ligule short (<1 mm); auricles large; upper sheath hairy or not; blade bright green, hairy, rough on margins. Woodlands .... ***Hordelymus europaeus***
Wood Barley

Bromopsis ramosa

A robust, tufted perennial up to 2 m tall. Leaves dark green, broad and hanging. Notable for the degree of hairiness of the leaf sheath which extends over the entire sheath. Leaf sheaths of *Bromus/Bromopsis* species are all tubular but split quite early on in the season to appear open, but they do not overlap the stem in the same way that a truly 'open' sheath does.

H A woodland species on damp, base-rich soils. Occurs in open woodland, wood-margins, hedgerows and formerly wooded roadsides, etc.

D Throughout the lowlands, scarce in upland Wales and Scotland.

S Native and stable.

Hordelymus europaeus

A short lived, loosely tufted perennial, up to 1 m tall. Differs from *Bromopsis ramosa* in generally shorter hairs that don't always extend to the upper sheath, narrower leaves, short ligule and larger auricles.

H Woodland and shady places on calcareous soils. Mainly lowland, especially in Beech woodlands. Old boundary banks and hedgerows.

D Lowland, mainly England.

S Declining due to loss of ancient woodbanks and hedgerows and coniferisation of deciduous woodlands.

16a. Annual; large spreading auricles (to 2 mm); leaves light green, weak; sheath inflated. Arable and waste places ....
.... ***Hordeum murinum***
Wall Barley

16b. Tufted perennials; auricles often obscure or absent; sheaths shortly hairy 17

17a. Auricles weakly developed (<0.5 mm); leaves narrow (1.5-5 mm), greyish green above, yellowish below, firm, the uppermost leaf standing erect at right angles; sheath tight fitting. Grassland .... ***Hordeum secalinum***
Meadow Barley

17b. Auricles indistinct with a purple collar; leaves wider (4-10 mm), pale or bright green; often with dark nodes on stem. Woodland .... ***Elymus caninus***
(*Agropyron caninum* •)
Bearded Couch

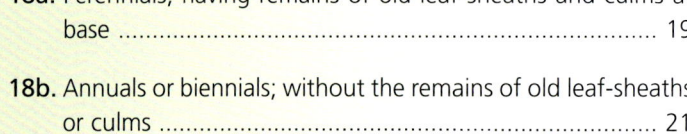

18a. Perennials; having remains of old leaf-sheaths and culms at base 19

18b. Annuals or biennials; without the remains of old leaf-sheaths or culms 21

Hordeum murinum

An annual species growing up to 60 cm tall. Sheaths hairy below, not hairy above; slightly inflated. Leaves are generally quite weak, they may be hairy. Notable for the large auricles which spread out from the sheath.

H Fertile waste ground and areas of disturbed soils; cultivated land and building sites, etc.

D Lowland. Scarce in Scotland, central Wales and NW England.

S Archaeophyte * (see Preston, Pearson and Dines 2002). (A species that became naturalised pre-1500.)

Hordeum secalinum

Differs from *H. murinum* in being perennial with much smaller auricles, closer fitting sheath and stiffer leaves. A good diagnostic characteristic is the angle of the upper most leaf which usually stands out at right angles to the shoot.

H Lowland grasslands on moderately heavy soils; coastal and inland meadows and pastures, especially on floodplains.

D Lowland. Concentrated SE of the Severn-Humber line with occassional records in Wales and NE England.

S Native and stable. Some decline possible from habitat loss through drainage and re-seeding.

Elymus caninus

Tussock forming with bright green, hairless leaves. With or without auricles the sheaths may be loosely hairy whilst a good diagnostic character is the dark nodes which are weakly hairy. Cf *Brachypodium sylvaticum* (a more common woodland grass), *E. caninus* has less dense tussocks and is much less hairy. Sparse hairs on sheaths are associated with the dark nodes whilst the leaves are duller green and much less conspicuously hairy.

H Locally common in woods and hedgerows, riverbanks and roadsides on freely drained, base-rich soils. Also occassionally in rocky habitats by the sea and mountain gullies. Common on roadsides.

D Widespread in England, Wales and S.Scotland.

S Native and stable.

19a. Auricles and junction of sheath and blade fringed on the margin with stiff hairs; leaf blades dull green, rough. Grasslands .... **Schedonorus arundinaceus**
(*Festuca arundinacea* • •)
Tall Fescue

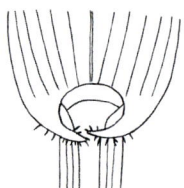

19b. Auricles and marginal junction of sheath and blade hairless; leaf blades bright green, and smooth beneath 20

20a. Leaf blades wide 6-16 mm, very glossy; ligule to 2.5 mm long; auricles pronounced. Woods and shady places .... **Schedonorus giganteus**
(*Festuca gigantea* • •)
Giant Fescue

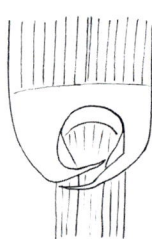

20b. Leaf blades narrow 3-8 mm; ligule short (to 1 mm long); auricles often obscure. Lowland grassland .... **Schedonorus pratensis**
(*Festuca pratensis* • •)
Meadow Fescue

Schedonorus arundinaceus

In *S. arundinaceus* the hairs on the auricles are diagnostic*. This is a very robust plant often forming dense tussocks with whitish lower sheath bases. Leaves tend to be dull with rough edges when rubbed downwards.

* Sterile hybrids with *Lolium perenne* and *Lolium multiflora* also have auricles with hairs but in the hybrids the leaves are glossy on one, or both, sides.

H Wide habitat range, particularly common in rough grassland, both inland in meadows and on coastal cliffs, also scrub and woodland margins. Different strains have different preferences; robust forms occur on heavy soils in low-lying meadows whilst smaller strains may occur on drier calcareous and sandy soils. Persists well in the absence of management and dense tussocks are a good indicator of tumble down meadows and pastures. A neutral grassland indicator.

D Widespread throughout except the higher mountains of Wales and Scotland.

S Native and stable. Possible increase since the 1960s may reflect better identification.

Schedonorus giganteus

A very distinct grass with glossy green leaves and very large auricles which clasp around the stem like a pair of pincers. The auricles are often reddish in colour. The leaves are bright green shiny and stiff. The ligule is longer than in *S. pratensis*. Distinct from the other tussock forming woodland grasses *Bromopsis ramosa* and *Brachypodium sylvaticum* in the lack of hairs on the sheath and leaves, and from *Brachypodium* in the presence of auricles.

H Damp woodlands and shady places on base-rich soils. Often with *Bromopsis ramosa* and *Brachypodium sylvaticum*.

D Widespread in the right habitat. Similar national distribution to *S. arundinaceus*.

S Native and stable.

Schedonorus pratensis

This is a more slender plant that rarely forms distinct tussocks. The auricles are generally smaller and less clasping than in *S. giganteus* and lack the hairs of *S. arundinaceus*. Leaves are not rough along their edges, and not as shiny as *S. giganteus*.

H Rich, moist soils in the lowlands, especially water meadows, pastures and roadsides.

D Widespread in areas where cultivation of meadows, etc., is possible.

S Native. Use in agricultural seed mixes has obscured the natural distribution of the species. Recent declines linked to loss of wet meadow habitats.

21a. Auricles very obscure, small; leaves blue-green and narrow (1-3.5 mm); sheath smooth. Saltmarshes .... ***Hordeum marinum***
Sea Barley

21b. Spreading auricles well developed; leaves wider (up to 10 mm). Waste places 22

22a. Blades glossy beneath, hairless; lower sheaths pink to red. Waste places .... ***Lolium multiflorum***
Italian Rye-grass

22b. Blades dull beneath, with short hairs; lower sheaths greenish or white, not pink. Waste places .... ***Hordeum murinum***
Wall Barley

No auricles (from Couplet 9) ▼

23a. Leaf-sheath open, with free margins, one margin overlapping the other (from couplet 9) 24

23b. Leaf-sheaths tubular, without overlapping margins 67

Hordeum marinum

A short annual species (up to 40 cm) with short, stiff, hairy leaves and an inflated sheath. Differs from the other *Hordeum* species which key out here in having very obscure auricles. A good diagnostic feature is the blue-green leaves.

H Coastal. Bare areas on margins of pools and grazing marshes, sea walls and tracks by the sea.

D Locally abundant in SE England and around the Severn estuary.

S Native. Probably declining along the south coast and north of the Wash due to sea wall defence construction and conversion of saltmarsh to arable land.

Lolium multiflorum

Youngest leaf rolled in the shoot is a diagnostic separator from *Lolium perenne* where youngest leaf is folded. They share the reddish pink basal leaf sheath colour but leaves of *L. multiflorum* tend to be wider. Differs from *Schedonorous pratensis* in having leaves which are glossy on the lower surface.

H Introduced for its forage qualities in the 1800s. Persists on roadsides, field margins and waste places.

D Widespread.

S Neophyte (introduced around 1830). May be expanding in Ireland.

Hordeum murinum

An annual species growing up to 60 cm tall. Sheaths hairy below, not hairy above, slightly inflated; notable for the large spreading auricles. Leaves are generally quite weak, they may be hairy.

H Fertile waste ground and areas of disturbed soils; cultivated land and building sites, etc.

D Lowland. Scarce in Scotland, central Wales and NW England.

S Archaeophyte * (see Preston, Pearson and Dines 2002). (A species that became naturalised pre-1500.)

Leaf-sheath open (from couplet 23) ▼

24a. Leaves narrow (2-6 mm), prominently ribbed on upper surface, ribs covered with minute hairs; perennial with long rhizomes. Coastal sandy areas 25

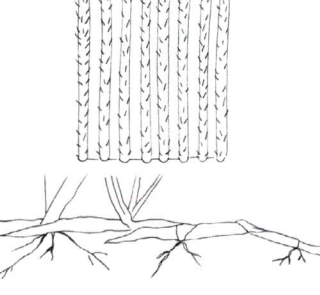

24b. Leaves may be narrow but lack ribs with hairs. Generally not coastal apart from some annual species 26

25a. Ligule long (1-3 cm); blade narrow (often inrolled when young or open to 6 mm wide), glaucus above, sometimes green below; rhizome stout. Sand dunes .... ***Ammophilla arenaria***
Marram Grass

25b. Ligule short (<1 mm); leaves inrolled and narrow, bluish grey; sheath white; rhizomes slender. Sandy beaches .... ***Elymus junceiformis***
(*Agropyron junciforme* • *Elymus farctus* • *Elytrigia juncea* •)
Sand Couch

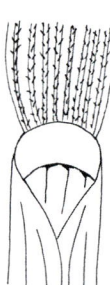

26a. Leaves aromatic (coumarin scented) when bruised; spreading hairs at the top of the sheath where it meets the blade; ligule 1-5 mm .... ***Anthoxanthum odoratum***
Sweet Vernal-grass

26b. Leaves and sheaths lack spreading hairs 27

Ammophila arenaria

A large grass forming dense patches through extensive rhizome growth. The extremely long ligules, up to 3 cm, are a good diagnostic feature. Leaves grey-green up to 90 cm long. Sheaths overlapping and leaf tips very sharp. Leaves tightly inrolled in the sheath but often open out when mature, but, at 6 mm width, are much narrower than those of *Leymus arenarius* with which it often grows on the most mobile dune systems.

🄷 Mobile sand dunes.

🄳 Throughout lowland coastal regions.

🄢 Native and stable. Attempts at inland colonisation on golf courses have generally failed.

Elymus junceiformis

Looks like a couch but contrasts with other species of *Elymus* in lacking auricles. Leaves blue-grey and narrow, sometimes spreading but often drooping, flat or often rolled. Careful observation with a hand lens will reveal hairs on the ribs of the upper surface of the leaves.

🄷 Loose sand and shingle on the foreshore. Its greater tolerance to salt in the air and soil than other sand dune species makes it one of the first colonisers of new dune formations.

🄳 Throughout lowland coastal regions.

🄢 Native and stable.

Anthoxanthum odoratum

Sheath smooth or just slightly hairy but forming a dense 'beard' of hairs at its top (where it meets the leaf blade). Lower blade slightly swollen which sometime gives a false impression of an auricle. Strongly scented with couramin especially when crushed or if weak then through taste. One of the first grasses to flower.

🄷 Abundant in a wide range of habitats; heaths, moors, pastures, meadows, open woodland. From heavy clay soils to light sandy soils.

🄳 Throughout.

🄢 Native and stable.

27a. Shoots thickened at ground level, the lowest 1-3 internodes swollen and more or less bulbous 28

27b. Shoots not bulbous or thickened at base 30

28a. Base of lower sheath yellow/orange at junction with shoot; basal internodes bulbous or pear shaped; upper leaf surface loosely hairy. Especially common on rough grassland and arable land .... ***Arrhenatherum elatius*** var ***bulbosus***
False Oat-grass

28b. Base of shoots and roots not yellow; basal internodes only slightly swollen; leaf not hairy, often grey-green 29

29a. Leaf blades wide (3-9 mm) and up to 45 cm long; ligule to 6 mm, obtuse; basal swelling often obscure ....
.... ***Phleum pratense pratense***
Timothy

29b. Leaf blades narrow (2-5 cm wide) and shorter (up to 12 cm long); ligule to 4 mm, acute; basal swelling usually pronounced
.... ***Phleum pratense bertolonii***
Smaller Cat's-tail

30a. Leaf blade hairless 31

30b. Leaf blade more or less hairy 57

Note: Some species which are variable in their leaf hairiness key out following either couplet here

Arrhenatherum elatius

Tall, loosely tufted with yellowish roots; var *bulbosus* has swollen lower internodes, with often 2 or 3 pear shaped swellings above the roots. Blades are coarse, often hairy above, or hairless, generally slightly rough.

🄷 Common in dry habitats including rough grassland, hedgerows, roadside verges, shingle and waste places. Intolerant of heavy grazing. Var. *bulbosus* is most common on road sides, and as a weed of arable ground on dry soils where it spreads readily through dispersal of the bulbils. A good indicator of neutral grassland.

🄳 Throughout, except parts of Highland Scotland.

🅂 Native and stable.

Phleum pratense pratense

Leaf blades soft, pale green; can give the impression of a small Dactylis but leaves are rolled not folded. Basal nodes usually swollen just below ground surface, round and either white or pink in contrast to *Arrhenatherum*'s pear shaped yellow bulbs. Easily detected by probing the soil around the base of the plant with your finger.

🄷 Grassy habitats; meadows, pastures, waysides. Tends to occur on damper, and heavier soils, than subsp. *bertolonii*. Widely sown in the past.

🄳 Throughout, but scarce in Highland Scotland.

🅂 Native and stable.

Phleum pratense bertolonii

Smaller than *P. pratense* with an acute rather than obtuse ligule.

🄷 Old meadows and pastures, generally on drier and lighter soils than subsp. *pratensis*.

🄳 Throughout lowland Britain; less common than subsp. *pratense* north of the central lowlands of Scotland.

🅂 Native and stable.

31a. Roots and bases of lower sheaths orange-yellow; leaf blades dull green, to 10 mm wide and 40 cm long, slightly rough on edges; ligule short 1-3 mm .... ***Arrhenatherum elatius***
False Oat-grass

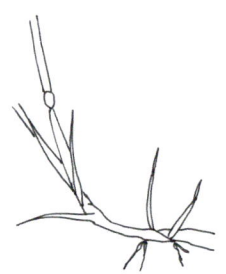

31b. Roots and bases of lower sheaths not yellow, or if so then leaves narrower and shorter 32

32a. Plants with creeping rhizomes, wiry or fleshy, white or brown; perennials 33

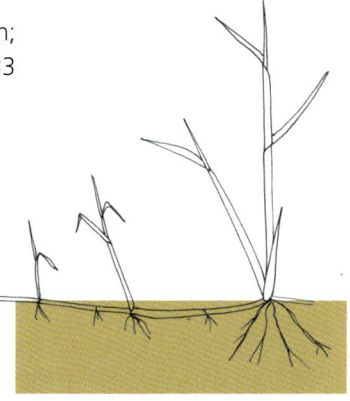

32b. Annuals, or perennials without rhizomes; may be loosely or densely tufted, or with creeping stoloniferous 39

Note: This is a very difficult dichotomy as the rhizomes are not always easy to see in a dense sward. The jizz of *Holcus mollis*, *Calamagrostis* and *Phalaris* is easy to learn, the two species of *Agrostis* are more difficult. Some of these species appear again further through the key if the rhizomes are missed.

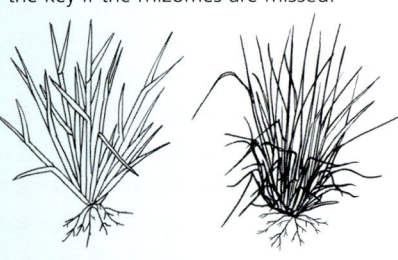
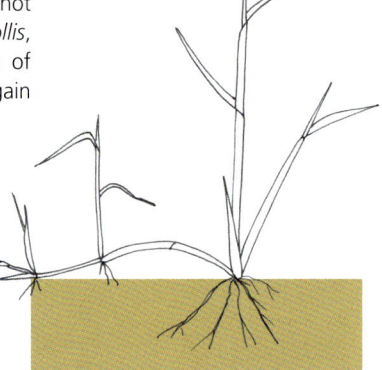

33a. Rhizomes well developed. Woodland and damp habitats 34

33b. Rhizomes often obscure, especially in dense swards. Dry grassland habitats 36

Arrhenatherum elatius

Tall, loosely tufted with yellowish roots; var. *bulbosus* has swollen lower internodes, with often 2 or 3 pear shaped swellings above the roots. Blades are coarse, often hairy above, or hairless, generally slightly rough.

H Common in dry habitats including rough grassland, hedgerows, roadside verges, shingle and waste places. Intolerant of heavy grazing. Var. *bulbosus* is most common on road sides and as a weed of arable ground on dry soils where it spreads readily through dispersal of the bulbils.

D Throughout, except parts of Highland Scotland.

S Native and stable.

34a. Leaves greyish green, may be sparsely hairy; nodes bearded; ligule 1-4 mm; rhizomes well developed. Woodland and hedge banks .... ***Holcus mollis***
Creeping Soft-grass

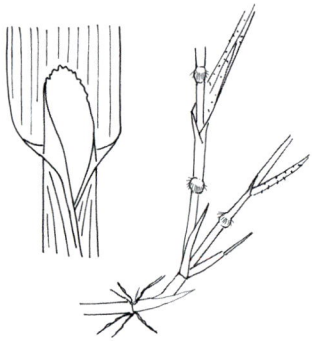

34b. Leaves not grey green in colour; ligule longer, up to 16 mm. Wet or damp places 35

35a. Blades closely ribbed on upper surface, up to 70 cm long and 4-10 mm wide, narrowed at base; rhizomes wiry. Damp woods and fens .... ***Calamagrostis epigejos***
Wood Small-reed

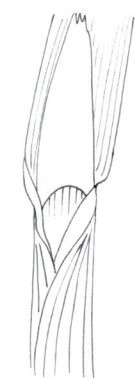

35b. Blades not ribbed but have cross-veins between nerves; leaves shorter (to 35 cm) and wider (6-18 mm wide), broader at the base; rhizomes fleshy. Wet places ..... ***Phalaris arundinacea***
Reed Canary-grass

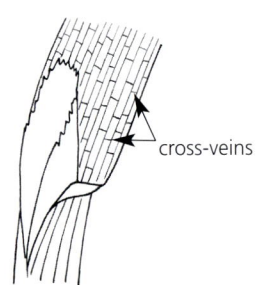

cross-veins

36a. Blades very narrow (1-2 mm), may be rolled and bristle like; ligules longer than wide, upper ones pointed; densely tufted. Acidic dry grassland and heath ..... ***Agrostis vinealis***
(*Agrostis canina montana* •)
Brown Bent

36b. Blades to 5 mm wide, flat; ligules shorter than wide, blunt or toothed 37

Holcus mollis

Forms extensive patches of widely spaced shoots. Leaves often pale, grey/green, may be slightly hairy. Sheaths smooth. Stems have distinct down pointing hairs at the nodes, otherwise hairless, but these may not be apparent early in the spring.

H Well drained soils, neutral to acidic; shady habitats including woodlands, hedgebanks, under bracken, heathlands. Occasional weed of sandy arable ground.

D Throughout.

S Native and stable.

Calamagrostis epigejos

A coarse grass with strong rhizomes forming loose patches of tall, narrow shoots. Leaves are generally dull, hairless and pointed with close veins on the upper surface.

H Open habitats on heavy soils; damp woodlands, ditches, fens. Occasional in ungrazed grasslands, coastal sand dunes and cliffs.

D Widespread in southern England; local in coastal areas of Scotland and Wales.

S Native and stable. Some increase since 1960 may represent relaxation of grazing, or better recording.

Phalaris arundinacea

Distinct from *Calamagrostis* in having cross veins in the leaved but lacks distinct ribs. Generally more robust and paler-green in colour.

H Damp habitats; ditches, Alder-Willow woodland, banks of rivers, lakes, etc. Tolerates greater soil water table fluctuation than Phragmites or *Glyceria maxima*. In the absence of cutting or grazing it may expand into damp grasslands to form extensive stands, palatable to grazing stock when young.

D Widespread. Local in Highland Scotland.

S Native and stable.

Agrostis vinealis

Rhizomes often obscure and shoots may appear clumped in dense grass turf. Clearer to see in heathland habitats where it tends to grow within subshrub patches in contrast to *A. capillaris* which forms dense grassy turfs. Leaves are narrow, often needle-like. Ligule is longer than wide and pointed. Easily separated from *A. canina* which has creeeping stolons and grows as a turf in open, wet, habitats.

H Sandy or peaty soils of heaths, grasslands, open woodlands.

D Widespread. Accurate distribution uncertain since often poorly separated from *A. canina* when they only had sub-species status.

S Native and stable.

37a. Ligule 0.5-2 mm, fringed with minute hairs; blade long, up to 45 cm, often inrolled, stiff and erect. Sheath may be sparsely hairy. An erect plant. Chalk or limestone grassland
.... **Brachypodium pinnatum**
Heath False-brome

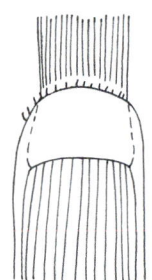

37b. Ligule not fringed with hairs; blades shorter (2-20 cm), not stiff; sheath glabrous. Not restricted to calcareous habitats
.... 38

38a. Ligule shorter than wide, 0.5-2 mm long; blades narrow, up to 5 mm wide; rhizome short. Grassland ....
.... **Agrostis capillaris**
(*Agrostis tenuis* •)
Common Bent

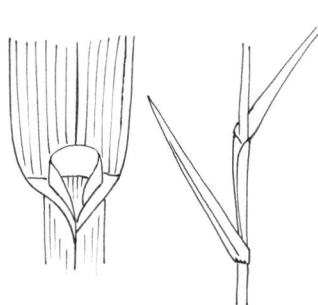

38b. Ligule as long as wide, 2-6 mm long, often toothed; blades wider (up to 8 mm); rhizomes long. Arable and waste ground
.... **Agrostis gigantea**
Black Bent

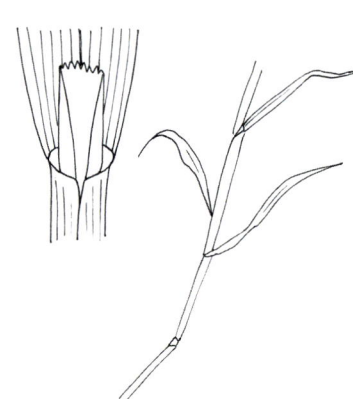

39a. Perennials; densely to loosely tufted or stoloniferous, having remains of old sheaths at base 40

39b. Annuals or biennials; lacking old sheaths at base; shoots solitary or clustered due to branching at the base 52

Brachypodium pinnatum

A tussock forming grass with stiff, narrow, leaves which are often tightly inrolled. Leaf blades may be sparsely hairy in common with *Bromus erectus* with which it often grows. However, the open sheath of *B. pinnatum* and lack of bristle-like hairs along the leaf margin should be sufficient to differentiate the two species. Calcareous grassland indicator.

H Dry, infertile, calcareous soils. Mainly grassland but also scrub, roadsides, quarries.

D Lowland. Restricted by incidence of chalk or limestone geology so principally in the central belt of England.

S Native, expanding due to reduced grazing pressure.

Agrostis capillaris

The dull leaves tend to spread at a distinct 45 degree angle to the stem. In contrast to the other species of *Agrostis* the ligule is shorter than it is wide. Forms dense turf in acid grasslands often with *Festuca ovina* and can be a minor component in other grasslands and heathlands.

H Wide tolerance of soil moisture and base status though tends to be most frequent on poor, neutral to slightly acidic substrates. Grassland, heathland, woodland and scrub, sand dunes, ruderal habitats.

D Common throughout. Sea level up to 1,200 m.

S Native and stable.

Agrostis gigantea

Larger than *A. capillaris* with a longer, often toothed ligule. The leaves are generally rather 'floppy' compared to the stiff, upright habit of *A. capillaris*.

H Rampant weed of neglected arable land and cornfields; especially on lighter soil. May form dense stands on wetter ground; in woodland, hedgerows, rough grassland.

D Lowland.

S Naturalised (pre 1500 AD). Increased since 1960s but still probably under-recorded. Often not distinguished from *A. stolonifera*.

40a. Leaf blades prominently grooved above, very rough on edges, smooth below, dark green; ligule very long (to 15 mm), pointed and stiff; plant densely tufted. Wet grasslands and woods .... ***Deschampsia cespitosa***
Tufted Hair-grass

40b. Leaf blades not grooved; ligules not stiff nor pointed and usually much shorter 41

41a. Plants lacking stolons; loosely to densely tufted; may have weak rhizomes 42

41b. Plants with creeping stolons, rooting at the nodes and there producing tufts of leafy shoots 50

42a. Leaf blades narrow (to 2.5 mm), and short, to 15 cm long ... 43

42b. Leaf blades wider (>2.5 mm and up to 15 mm), and longer, up to 40 cm long 45

43a. Ligule longer than wide (1-5 mm); leaves uniformly green/grey green. Dry acidic heath and grassland habitats ....
.... ***Agrostis vinealis***
(*Agrostis canina montana* •)
Brown Bent

43b. Ligule shorter than wide (0.5-2.0 mm). Dry neutral grassland
.... 44

Deschampsia cespitosa

A very tough and rough leaved grass – you risk cutting your fingers if you rub them backwards down the leaf blade, often a good way to identify its presence in a dense grass sward! The ligule is long and pointed and leaves deep green, often slightly shiny. Neutral grassland indicator.

🄷 Wet and poorly drained soils; marshy grassland, moorland and woodland. Forms large tussocks in the absence of cutting or grazing.

🄳 From sea level to 1,200 m in Scotland.

🅂 Native and stable.

Agrostis vinealis

Rhizomes often obscure and shoots may appear clumped. Clearer to see in heathland habitats where it tends to grow within subshrub patches in contrast to *A. capillaris* which forms dense grassy turfs. Leaves are narrow, often needle-like. Ligule is longer than wide and pointed.

🄷 Sandy or peaty soils of heaths, grasslands, open woodlands.

🄳 Widespread. Accurate distribution uncertain since often poorly separated from *A. canina* when only sub-species status.

🅂 Native and stable.

44a. Blades glossy beneath, dark green, often crimped between 1/2-3/4 from the base; basal sheath often streaked with bright yellow, yellowish brown (nicotine) when old ....
.... ***Cynosurus cristatus***
Crested Dog's-tail

44b. Blades dull beneath, pale green, often sparsely hairy; basal leaf-sheath whitish, hairy with downward pointing hairs ......
.... ***Trisetum flavescens***
Yellow Oat-grass

45a. Ligules long (3-10 mm). Woodland and disturbed habitats ... 46

45b. Ligules short (to 2.5 mm) 47

46a. Leaves wide (5-15 mm); ligule long 2-4 (-10 mm); blades thin, dull green above, shiny below; tufted. Woodland ....
.... ***Milium effusum***
Wood Millet

46b. Leaves narrower (2-8 mm); ligule shorter (1.5-6.0 mm); leaves dull green on both sides. Arable and waste places ....
.... ***Agrostis gigantea***
Black Bent

Cynosurus cristatus

A rather anonymous grass, often overlooked when not in flower. The folded leaf expression may be missed for *Lolium perenne* but *Cynosurus* leaves are only shiny on the lower surface, they lack auricles and the sheath base tends to be nicotine yellow rather than the red of *Lolium*. Another useful character is the 'pinch' point about 2/3 up the leaf. Neutral grassland indicator.

H Grasslands, especially pastures and hay meadows. Widespread but avoids extremes of soil water logging, drought and pH.

D Throughout; generally lowland but up to 850 m in Westmoreland.

S Native and stable.

Trisetum flavescens

The downward pointing hairs on the leaf sheath are a good diagnostic character in the field. The leaves are generally delicate and sparsely hairy above, dull below.

H Grasslands; especially on calcareous, freely drained soils. Susceptible to over-grazing.

D Lowland, 0-550 m.

S Native and stable.

Milium effusum

A loosely tufted perennial with broad leaves and a long ligule. The leaves are shiny above and dull below, floppy rather than stiff. More or less restricted to woodland habitats.

H Damp, calcareous soils; especially oak and beech woodlands and rocky coastal areas in NW Scotland.

D Generally lowland, 0-350 m. Absent from Highland Scotland.

S Native and stable.

Agrostis gigantea

Larger than *A. capillaris* with a longer, often toothed ligule. The leaves are generally rather 'floppy' compared to the stiff, upright habit of *A. capillaris*.

H Rampant weed of neglected arable land and cornfields; especially on lighter soil. May form dense stands on wetter ground; in woodland, hedgerows, rough grassland.

D Lowland.

S Naturalised (pre 1500 AD). Increased since 1960s but still probably under-recorded. Often not distinguished from *A. stolonifera*.

47a. Leaves blade narrow, generally <5 mm wide 48

47b. Leaves blade generally wider, 3-13 mm; basal sheaths often purple or whitish 49

48a. Ligule 0.5-2 mm, fringed with minute hairs; blades narrow, often tightly inrolled, up to 45 cm long, erect and stiff; leaves and sheaths may be sparsely hairy; tufted, erect plant. Chalk and limestone grassland .... **Brachypodium pinnatum**
Heath False-brome

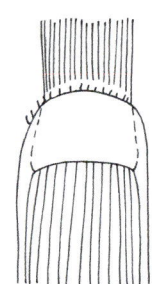

48b. Ligule not fringed with hairs; blades shorter (2-20 cm), flat, usually at an angle of 45 degrees from the stem; plant entirely glabrous. Wide ranging in grassland habitats .... **Agrostis capillaris**
(*Agrostis tenuis* •)
Common Bent

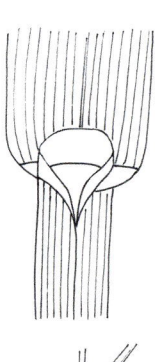

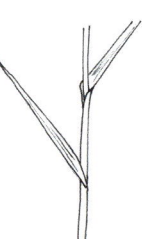

Brachypodium pinnatum

A tussock forming grass with stiff, narrow, leaves which are often tightly inrolled. Leaf blades may be sparsely hairy in common with *Bromopsis erecta* with which it often grows. However, the open sheath of *B. pinnatum* and lack of bristle-like hairs along the leaf margin should be sufficient to differentiate the two species.

H Dry, infertile, calcareous soils. Mainly grassland but also scrub, roadsides, quarries.

D Lowland. Restricted by incidence of chalk or limestone geology so principally in the central belt of England.

S Native, expanding due to reduced grazing pressure.

Agrostis capillaris

The dull leaves tend to spread at a distinct 45 degree angle. In contrast to the other species of *Agrostis* the ligule is shorter than it is wide. Forms dense turf in acid grasslands and can be a minor component in other grasslands and heathlands.

H Wide tolerance of soil moisture and base status though tends to be most frequent on poor, neutral to slightly acidic substrates. Grassland, heathland, woodland and scrub, sand dunes, ruderal habitats.

D Common throughout. Sea level up to 1,200 m.

S Native and stable.

49a. Basal sheaths purplish, glabrous; ligules 1 to 2.5 mm; leaf blade asymmetric at junction with sheath, glabrous, dark green. Meadow species; common .... ***Alopecurus pratensis***
Meadow Foxtail

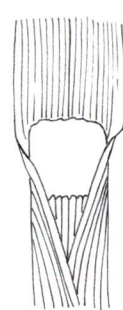

49b. Basal sheaths greenish or whitish, lower often slightly hairy near the nodes; ligules very short (0.5-1.5 mm); leaves may be loosely hairy above, pale or bright green; dark nodes on stem are distinctive. Woods, shaded places .... ***Elymus caninus***
(*Agropyron caninum* •)
Bearded Couch

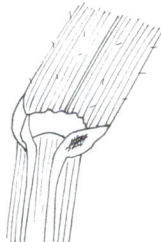

50a. Blades very narrow (1-2 mm), bright green, soft, usually forming a fine turf; ligule longer than wide (1-4 mm), pointed, entire; stolons with compact tufts of leaves at nodes. Damp grassland .... ***Agrostis canina***
(*Agrostis canina canina* •)
Velvet Bent

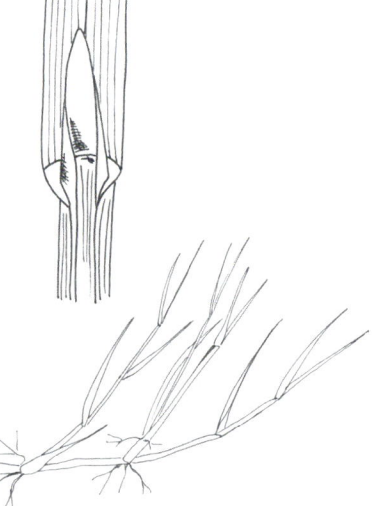

50b. Blades wider, dull or greyish green; ligules blunt 51

Alopecurus pratensis

Rather anonymous. In early spring conspicuous with leaves that are wider, duller and darker than other meadow grasses. A good field character is the asymmetry of the junction of the leaf blade with the sheath. Otherwise the leaves lack hairs, the ligule is short and the basal sheath is often tinted purple.

H Fertile grasslands on damp but not waterlogged soils; especially floodplain meadows. Also roadsides and woodland margins.

D Lowland, throughout England, to 610 m in Dumfrieshire. Scarce in the north and western Highlands of Scotland.

S Native and stable.

Elymus caninus

Loosely tufted woodland grass. Similar to *A. pratensis* but lacks the leaf sheath asymmetry. Dark nodes on the stem are a good diagnostic. Basal sheaths are more likely to be white or green rather than purple. Leaves are bright green and hairless; the lower sheaths of some plants may be loosely hairy. Auricles, if present, are poorly developed. These two species are unlikely to overlap due to their different habitat requirements.

H Locally common in woods and hedgerows, riverbanks and roadsides on freely drained, base-rich soils. Also occassionally in rocky habitats by the sea and mountain gullies.

D Widespread in England, Wales and southern Scotland.

S Native and stable.

Agrostis canina

One of a trio of stoloniferous grasses of damp habitats. The leaf blade is narrower than *Agrostis stolonifera* and *Alopecurus geniculatus* and generally a brighter green colour. Forms 'soft' turf of compactly growing plants lacking the pink sheath of *A. stolonifera* and the waxy white of the *A. geniculatus*. The ligule is usually pointed.

H Generally infertile habitats; permanent grasslands, heaths, mires and flushes on acidic peaty substrates, also open habitats on mineral soils which are permanently damp including tracks, ditches, water margins, etc.

D Throughout but scarce in the Midlands, north east England and the Highlands.

S Native. Probably stable but many earlier records for the aggregate make precise determination of change difficult.

51a. Lower leaf-sheaths green or purplish; ligule blunt, often tattered; forming a close turf. Grassland, common .... ***Agrostis stolonifera***
Creeping Bent

51b. Lower leaf-sheaths usually having a whitish waxy covering, upper sheath inflated, nodes may be purplish; ligule blunt, entire; shoots bend at lower joints; not turf-forming. Wet places .... ***Alopecurus geniculatus***
Marsh Foxtail

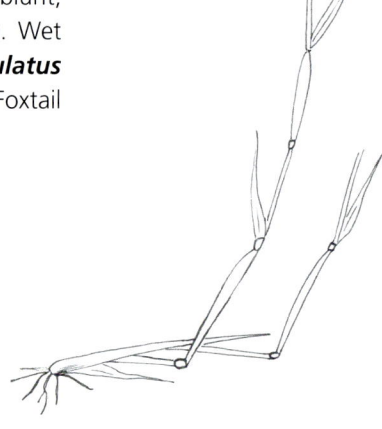

Annuals (from Couplet 39) ▼

52a. Small, stiff, plants, up to 40 cm tall. Leaves <10cm long and <4 mm wide 53

52b. Larger plants, up to 80 cm tall. Leaves longer (up to 25 cm) and wider (up to 10 mm wide) 56

Agrostis stolonifera

Compared to *A. canina* the leaves are broader, the ligule blunter and often tattered, and the sheaths often have a purple tint. Individual shoots tend to be more discernable than in *A. canina* and form less dense carpets. Often a component of damp meadows and pastures where the stoloniferous character is hard to see. See *Alopecurus geniculatus* for differences.

H Wide range of habitats but generally those where soils are damp, at least for part of the year. Permanent neutral grasslands, saltmarshes, dune slacks, cliff tops, and flushed areas including springs, ditches and streamsides. Occasionally found in arable fields and woodlands.

D Throughout.

S Native and stable.

Alopecurus geniculatus

Similar in stature to *Agrostis stolonifera* but lower sheaths lack purple tint being generally green or white, whilst nodes are often deep purple. Upper sheath inflated. Stems articulate at the nodes more strongly than A. stolonifera when that is growing in open habitats.

H Fertile, winter flooded grasslands, pond edges, ditches. Often colonising areas of bare mud and pool edges it can also occur in damp grasslands. Readily colonises areas of bare and compacted ground. Mainly on circum neutral soils.

D Generally lowland but up to 595 m in Northumberland. More widespread in the north of Scotland than *A. pratensis*.

S Native and stable.

53a. Leaves pale green/whitish, only 0.5-6.0 cm long, <4 mm wide, smooth on upper surface; ligule long, up to 7 mm; sheath inflated .... ***Phleum arenarium***
Sand Cat's-tail

53b. Leaves green/grey green, up to 10 cm long, blade rough above; ligule shorter (<3 mm) 54

54a. Leaf blades pointed, dark grey/green; ligule 0.3-1 mm. Damp coastal habitats, mainly saltmarshes .... ***Parapholis strigosa***
Hard-grass

54b. Ligule longer, 0.5-3 mm. Dry habitats, often by the sea 55

55a. Leaf blade green or purplish, finely nerved (not ribbed), leaf tip pointed, narrow (0.5-2 mm); leaf blade incurved and sickle-shaped. [Panicle one sided, branched at the base, up to 25 mm wide.] .... ***Catapodium rigidum***
(*Desmazeria rigida* •)
Fern-grass

55b. Leaf blade dark green, fleshier, ribbed above, blunt tipped, wider (1-3.5 mm). [Panicle narrower (up to 12 mm wide) and less branched.] .... ***Catapodium marinum***
(*Desmazeria marina* •)
Sea Fern-grass

Comment: this is a difficult separation when not flowering.

Phleum arenarium

The first of a series of winter annuals; germinating in late summer/autumn and over-wintering as a small plant. Very short with inflated upper sheaths. Leaves are pale, hairless and generally less stiff than those of the *Catapodium* species.

🄷 Coastal, dry habitats especially sand dunes where it occurs with *Ammophila arenaria* and other annual grass species. Also, sandy shingle. Occasionally on dry heaths and other sandy places.

🄳 Coastal, locally common except in northern Scotland. Inland in Breckland and occassionally elsewhere where sand has been imported from coastal habitats.

🅂 Native, stable but declining in Breckland.

Parapholis strigosa

The prostrate growth form helps distinguish *Parapholis* from the other annuals. Its grey colour and small size help distinguish it from other creeping grasses in the same habitat. A very distinct grass when in flower but otherwise easily overlooked in the dense grazed swards of *Festuca rubra-Agrostis stolonifera* in which it often occurs.

🄷 Damp, often brackish habitats especially salt marshes and saltmarsh-sand dune transitional zones that are kept open by grazing. Also occurs on muddy banks, shingle ridges and sea walls.

🄳 Coastal, throughout England, Wales and north to Mull and mid-Lothian. Generally scarce in the north of Scotland and Northern Ireland.

🅂 Native and stable.

Catapodium rigidum

Another winter annual. Leaves are narrower and pointed compared to *C. marinum* but a good diagnostic character is the incurved leaf blade which curves like a sythe. The panicle is more obviously branched.

🄷 Infertile, bare substrates in open, coastal areas including dry banks, cliff tops, sand dunes, and on walls, pavements, etc. Often occurs with other annual dune grasses.

🄳 Lowland, mainly coastal, especially in the south and west of England, Wales and Ireland. Scarce in NE England and NE Scotland. Inland along salt treated roadsides.

🅂 Native and stable. Localised expansion in the SE of England along roadsides in the M4 corridor.

Catapodium marinum

Overwintering as small plants. Generally more robust than *C. rigidum* and the blunt leaf tip gives the appearance of a small *Poa* but the leaves here are rolled not folded. In flower the narrower, unbranched panicle is a good character.

🄷 Dry, open habitats, often on calcareous substrates.

🄳 Lowland, throughout southern England, coastal Wales and Northern Ireland. Scarce in Scotland where it reaches its world northern limit at the Forth of Tay.

🅂 Native and stable.

56a. Leaf blades prominently ribbed, rough, 3-10 mm wide and up to 25 cm long; ligule 3-10 mm, oblong. Arable and waste places, generally dry and sandy .... **Apera spica-venti**
Loose Silky-bent

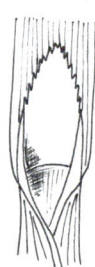

56b. Leaf blades only finely ribbed, blade narrow (2-8 mm wide) and only up to 16 cm long; ligule shorter (2-5 mm), blunt; sheaths slightly inflated, green/purple. Arable and waste places .... **Alopecurus myosuroides**
Black-grass

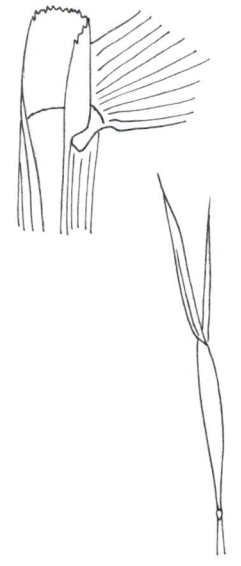

Sheaths open leaves usually hairy (from Couplet 30) ▼

57a. Roots and bases of lower sheaths orange yellow at junction with shoot; sheaths hairless; leaves loosely or sparsely hairy, 4-10 mm wide, dull green. Very common species of rough grassland .... **Arrhenatherum elatius**
False Oat-grass

57b. Roots and leaf sheaths not so coloured or, if with yellowish roots, then having narrower leaves 58

Apera spica-venti

A winter annual that germinates in both the autumn and spring. Plants have the general appearance of *Agrostis capillaris*, but in *Apera* the ligule is longer than broad and leaves are generally wider with ribs on the upper surface.

H Lowland, open and disturbed sandy or light loam soils. A weed of arable crops in south, east and SE England; elsewhere occurs occasionally on waste ground. Can form dense populations.

D Generally concentrated in south and east England with older records in the NE and SW. Rare in Wales and Scotland.

S Archaeophyte but widely distributed outwith its Eurosiberian-Boreal temperate home range. Some consider it native in Britain with the first known records being from the 17th Century.

Alopecurus myosuroides

Characterised by slightly inflated, purplish sheaths and blunt ligule.

H Arable land and neglected grasslands. A frequent weed of cereal crops, especially those sown with wide row spacing. Favours damp, compacted soils.

D Lowland, especially in the SE; and south of the Humber-Severn estuary line.

S Archaeophyte though common outwith its Southern-Temperate European zone. Stable or slightly increasing due to changes in cropping methods.

Arrhenatherum elatius

Tall, loosely tufted with yellowish roots; var. *bulbosus* has swollen lower internodes, with often 2 or 3 pear shaped swellings above the roots. Blades are coarse, often hairy above, or hairless, generally slightly rough.

H Common in dry habitats including rough grassland, hedgerows, roadside verges, shingle and waste places. Intolerant of heavy grazing. Var. *bulbosus* is most common on road sides and as a weed of arable ground on dry soils where it spreads readily through dispersal of the bulbils.

D Throughout, except parts of Highland Scotland.

S Native and stable.

58a. Sheaths hairless, smooth; blades glossy beneath 59

58b. Sheaths loosely to densely hairy 60

59a. Ligule 2-5mm, rounded; rhizomes long forming loose tufts or patches. Fens and swampy places ....
.... ***Calamagrostis canescens***
Purple Small-reed

59b. Ligule shorter, 0.5-1.5mm, blunt; rhizomes short forming compact tufts. Grasslands .... ***Cynosurus cristatus***
Crested Dog's-tail

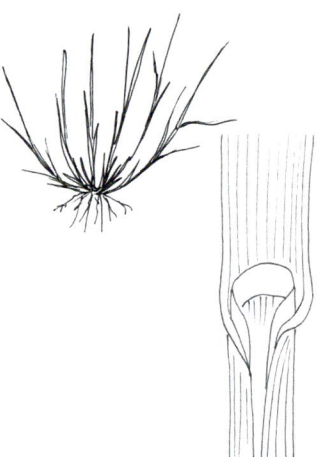

60a. Basal leaf sheaths white or purplish having pink veins (pyjama stripes), sheaths densely and velvety hairy. Very common .... ***Holcus lanatus***
Yorkshire Fog

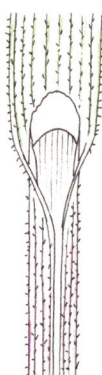

60b. Sheaths hairy but lack purple stripes 61

Calamagrostis canescens

Sheaths hairless and leaves sparsely hairy above. Contrasts with the more robust *C. epigejos* which has hairless leaves and a longer, ragged ligule.

🄷 Damp, often species-rich fen meadows and marshes. Also Alder and Willow Carr. Grows well in light shade and might become locally abundant.

🄳 Lowland. Core area is central and eastern England (East Anglia, Lincolnshire, south Yorkshire and into the north Midlands). Scattered in Scottish borders and Hampshire. Absent from Ireland.

🅂 Native. Declining due to habitat loss through falling water tables and natural succession due to neglect of water margin habitat.

John Hanley

Cynosurus cristatus

A rather anonymous grass, often overlooked when not in flower. The folded leaf expression may be missed for *Lolium perenne* but *Cynosurus* leaves are only shiny on the lower surface, they lack auricles and the sheath base tends to be nicotine yellow rather than the red of *Lolium*. Another useful character is the 'pinch' point about 2/3 up the leaf.

🄷 Grasslands, especially pastures and hay meadows. Widespread but avoids extremes of soil water logging, drought and pH.

🄳 Throughout; generally lowland but up to 850 m in Westmoreland.

🅂 Native and stable.

Holcus lanatus

A distinctive grass due to the purple vertical stripes on the lower leaf sheaths giving the grass the common descriptor of having 'pyjamas'. In contrast to *H. mollis* the hairs on the shoots are continuous and not restricted to the nodes. Also *H. lanatus* is not rhizomatous and can form dense clumps in open or unmanaged habitats; elsewhere it is a common component of grasslands on moist but not waterlogged soils across a wide range of habitats.

🄷 A very cosmopolitan species across most habitats. It lacks strong preferences for soil moisture, base status or fertility though is most favoured on neutral, moist soils that are not prone to extended waterlogging.

🄳 Throughout.

🅂 Native and stable.

61a. Plants with creeping, wiry rhizomes 62

61b. Plants without rhizomes 63

62a. Leaves sparsely hairy or hairless, grey green; ligule up to 5mm; sheath smooth, but nodes densely bearded. Shady places and arable land .... ***Holcus mollis***
Creeping Soft-grass

62b. Leaves sparsely hairy, green or yellow-green; ligule shorter, 1-2 mm, fringed with minute hairs. Chalk and limestone .... ***Brachypodium pinnatum***
Heath False-brome

63a. Leaf blades narrow (1-4 mm); sheath softly and densely hairy 64

63b. Leaf blades 4-15 mm or wider; sheath loosely or sparsely hairy 65

Holcus mollis

Strongly rhizomatous forming extensive patches of loosely spaced shoots. Leaves often pale or grey green. Young shoots smooth, mature stems have distinct beards of down pointing hairs at the nodes but are smooth between nodes.

H Well drained soils, neutral to acidic; shady habitats including woodlands, hedgebanks, under bracken, heathlands. Occasional weed of sandy arable ground.

D Throughout.

S Native and stable.

Brachypodium pinnatum

A tussock forming grass with stiff, narrow, leaves which are often tightly inrolled. Leaf blades may be sparsely hairy in common with *Bromopsis erecta* with which it often grows. However, the open sheath of *B. pinnatum* and lack of bristle-like hairs along the leaf margin should be sufficient to differentiate the two species.

H Dry, infertile, calcareous soils. Mainly grassland but also scrub, roadsides, quarries.

D Lowland. Restricted by incidence of chalk or limestone geology so principally in the central belt of England.

S Native, expanding due to reduced grazing pressure.

64a. Compactly tufted; ligule <1 mm; blades green or grey green, often inrolled and bristle-like, (1.0-2.5 mm wide), strongly ribbed on upper surface; hairs on sheath spreading. Dry often calcareous soils .... **Koeleria macrantha**
(*Koeleria cristata* •)
Crested Hair-grass

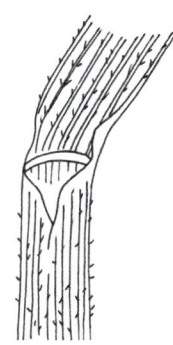

64b. Loosely tufted; ligule 0.5-2 mm; blades flat, green, wider (2.0-4.0 mm); lower sheath hairs downward pointing. Grasslands .... **Trisetum flavescens**
Yellow Oat-grass

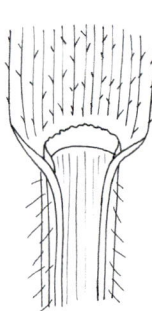

65a. Plants with solitary shoots; ligule blunt up to 6 mm; blade hairy on margins; annual .... **Avena fatua**
Wild-oat

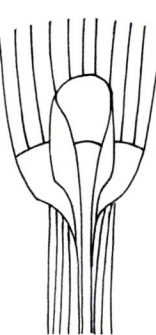

65b. Plants tufted, perennial, with remains of old sheaths and culms at the base; blades hairy above 66

Koeleria macrantha

A slender, often stiff grass, with a densely hairy sheath – the hairs spreading outwards in contrast to the down-pointing hairs on *Trisetum*. Very variable in its growth form and leaf characteristics such that it also keys out in the 'needle leaf' section where leaves on non-flowering stems are often in-rolled and <1 mm wide. The strongly ribbed leaves may be glaucous or green, glabrous or hairy.

🄷 Dry sandy areas near the sea, including sand dunes and cliff top grasslands. Also in open habitats on calcareous substrates or freely drained shallow soils of low fertility; in pastures and also on rocky  outcrops, scree slopes and old mining spoil.

🄳 Generally lowland and coastal. Inland distribution is concentrated on the chalk and limestone of central England and Yorkshire.

🅂 Native and stable. Some recent decline due to habitat loss.

Trisetum flavescens

The downward pointing hairs on the leaf sheath are a good diagnostic character in the field. The leaves are generally delicate and sparsely hairy above, dull below.

🄷 Grasslands; especially on calcareous, freely drained soils. Susceptible to over-grazing.

🄳 Lowland, 0-550 m.

🅂 Native and stable.

Avena fatua

Annual with solitary stout shoots and a long blunt ligule.

🄷 Cultivated land, but also on roadsides and waste places.

🄳 Concentrated in England where arable farming predominates. Elsewhere more or less restricted to the central lowlands of Scotland and Aberdeenshire, and SW Wales.

🅂 Archaeophyte of Mediterranean origin, but now widely distributed outside of this region.

66a. Lower leaf-sheaths with conspicuous spreading or reflect hairs; ligule long (to 6 mm). Woodland habitats ....
.... ***Brachypodium sylvaticum***
False Brome

66b. Lower leaf-sheath only slightly hairy around the nodes; ligule short (0.5-1.5 mm); sometimes with obscure auricles. Shady places .... ***Elymus caninus***
(*Agropyron caninum* •)
Bearded Couch

Leaves rolled, leaf sheath tubular (without overlapping margins) (from Couplet 23)

67a. Leaf-sheaths producing a slender bristle at apex, 1-4 mm long, on the side opposite the blade. Woodland ....
.... ***Melica uniflora***
Wood Melick

67b. Not as above 68

68a. Leaf-sheaths hairless 69

68b. Leaf-sheaths hairy 72

Brachypodium sylvaticum

A conspicuously hairy, bright green grass, with dense, reflexed, hairs on the leaf sheath throughout the plant and on its leaves. Tends to form tussocks.

🅗 A grass of woodland, scrub and shady hedgebanks. Generally on relatively freely drained neutral to calcareous soils. May also persist in the open following tree clearance, and occasionally on limestone pavement, screes and cliffs.

🅓 Throughout on suitable substrates but absent from much of NE Scotland.

🅢 Native and stable.

Elymus caninus

Generally much less hairy than *Brachypodium sylvaticum* with hairs restricted to lower sheath adjacent to the nodes. Leaves also much less conspicuously hairy and much duller green compared to *B. sylvaticum*. Tussocks less dense. A distinctive feature is the dark nodes on the stem in mature plants. May have indistinct auricles.

🅗 Locally common in woods and hedgerows, riverbanks and roadsides on freely drained, base-rich soils. Also occassionally in rocky habitats by the sea and mountain gullies.

🅓 Widespread in England, Wales and S.Scotland.

🅢 Native and stable.

Melica uniflora

Unique in the presence of a sharp bristle arising from the top of the leaf sheath and opposing the ligule. Leaves are narrow, bright green, and slightly 'floppy'. It forms extensive patches due to short rhizomatous growth. Sheaths may have sparse reflex hairs or be hairless.

🅗 Freely drained, usually basic, soils in slight to moderate shade including woodland rides and clearings, hedgebanks, roadsides and occasionally on rock ledges.

🅓 Widespread but scarce in northern Scotland, the southern uplands and parts of eastern England (Fenlands and Lincolnshire).

🅢 Native and stable.

69a. Annual with solitary shoots, branching at base; leaves loosely hairy; ligule short, toothed .... ***Bromus secalinus***
Rye Brome

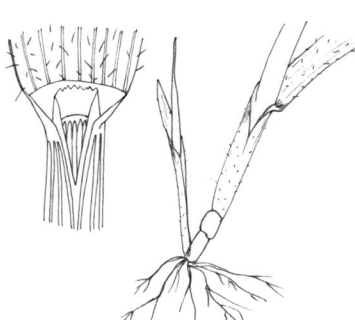

69b. Perennial, blades usually hairless 70

70a. Blades 3-10 mm wide and up to 40 cm long; basal sheaths whitish or purplish .... ***Alopecurus pratensis***
Meadow Foxtail

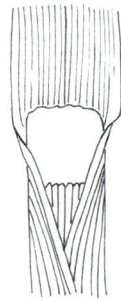

70b. Blades narrower (<4 mm) and shorter (<15 cm); basal sheath white or yellow 71

71a. Blades 1-4 mm wide, glossy beneath often crimped 1/2-3/4 from base; basal sheaths often streaked with yellow near junction with shoot .... ***Cynosurus cristatus***
Crested Dog's-tail

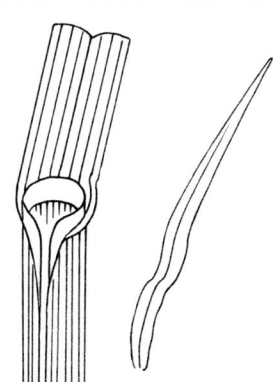

71b. Blades 2-4 mm wide, often twisted, with thickened, blunt, point; basal sheaths white, sometimes becoming yellowish brown .... ***Briza media***
Quaking-grass

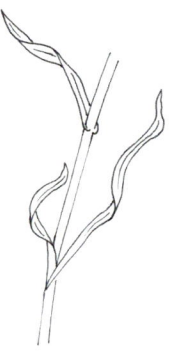

Bromus secalinus

An annual with solitary shoots, may occasionally function as a biennial. The sheaths are generally hairless but leaves are loosely hairy. Ligules are short (1-2 mm) and slightly toothed.

Ⓗ Mainly arable, especially cereals and waste ground. Occasionally in short term grass leys.

Ⓓ Lowland, currently more or less confined to SE England, the SW peninsula and SW Midlands.

Ⓢ Archaeophyte. Declined since the 19th and 20th Century. Recent recovery in Norfolk and Worcestershire.

Alopecurus pratensis

Rather anonymous. In early spring conspicuous with leaves that are wider, duller and darker than other meadow grasses. A good field character is the asymmetry of the junction of the leaf blade with the sheath. Otherwise the leaves lack hairs, the ligule is short and the basal sheath is often tinted purple.

Ⓗ Fertile grasslands on damp but not waterlogged soils; especially floodplain meadows. Also roadsides and woodland margins.

Ⓓ Lowland, throughout England, to 610 m in Dumfrieshire. Scarce in the north and western Highlands of Scotland.

Ⓢ Native and stable.

Cynosurus cristatus

Another rather anonymous grass, often overlooked when not in flower. The folded leaf expression may be missed for *Lolium perenne* but *Cynosurus* leaves are only shiny on the lower surface, they lack auricles and the sheath base tends to be nicotine yellow rather than the red of *Lolium*. Another useful character is the 'pinch' point about 2/3 up the leaf.

Ⓗ Grasslands, especially pastures and hay meadows. Widespread but avoids extremes of soil water logging, drought and pH.

Ⓓ Throughout; generally lowland but up to 850 m in Westmoreland.

Ⓢ Native and stable.

Briza media

A very distinctive grass when flowering, *Briza* can be hard to pick out when growing vegetatively within a sward. The leaves are winter green, short and tend to be slightly twisted. The ligule is short and blunt.

Ⓗ Meadows and pastures on generally well drained neutral to calcareous substrates. Also a component of calcareous fens, acidic flushes and dune slack grasslands.

Ⓓ Widespread but scarce in west Wales, the SW Peninsula and the north and west of Scotland.

Ⓢ Native and stable. Some recent declines due to landuse change and loss of lowland grasslands.

72a. Plants with rhizomes; leaves sparsely hairy, grey-green; nodes conspicuously bearded .... ***Holcus mollis***
Creeping Soft-grass

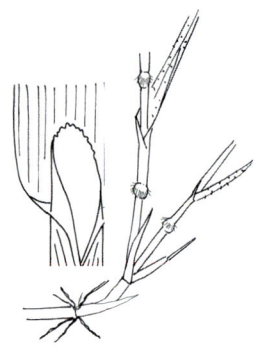

72b. Plants without rhizomes 73

73a. Perennials; plants tufted with remains of old sheaths at the base 74

73b. Annuals; solitary shoots but sometimes branched at the base 75

74a. Basal leaf-sheaths with dark, conspicuous, pink veins (pyjama stripes); leaves softly and densely hairy, blades 3-10 mm wide, soft. Very common .... ***Holcus lanatus***
Yorkshire Fog

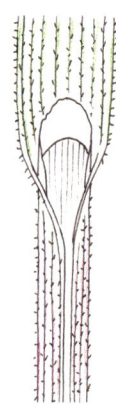

74b. Basal leaf-sheaths with greenish veins; lower leaf blades narrow, inrolled, 2-3 mm wide when opened, with stiff bristles on margins like a fish-bone. Calcareous grassland, locally common .... ***Bromopsis erecta***
(*Bromus erectus* • •)
Upright Brome

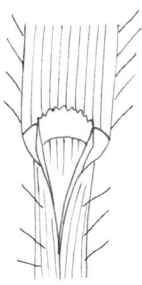

Holcus mollis

Strongly rhizomatous forming extensive patches of loosely spaced shoots. Often pale, or grey green, in colour. Stems have distinct 'beards' of down pointing hairs at the nodes. Sheaths between nodes not hairy. Beards are often not apparent early in the spring.

H Well drained soils, neutral to acidic; shady habitats including woodlands, hedgebanks, under bracken, heathlands. Occasional weed of sandy arable ground.

D Throughout.

S Native and stable.

Holcus lanatus

A distinctive grass due to the purple vertical stripes on the lower leaf sheaths giving the grass the common descriptor of stripy 'pyjamas'. In contrast to *H. mollis* the leaves are more densely hairy and the hairs on the stems are continuous, and not restricted to the nodes. Also *H. lanatus* is not rhizomatous and can form dense clumps in open or unmanaged habitats; elsewhere it is a common component of grasslands on moist but not waterlogged soils across many habitats.

H A very cosmopolitan species across most habitats. It lacks strong preferences for soil moisture, base status or fertility though is most favoured on neutral, moist soils that are not prone to extended waterlogging.

D Throughout.

S Native and stable.

Bromopsis erecta

A densely tufted, winter green, perennial. Lower leaves tend to be narrow and inrolled but upper leaves, when flattened, have very distinct bristles along their margins which arise at c.45 degrees to the leaf edge giving a 'fish bone' appearance. This character helps distinguish it from *Brachypodium pinnatum* which grows in similar habitats and also has a hairy sheath and narrow, often inrolled, leaves.

H Infertile, calcareous soils on chalk and limestone where it may become dominant in the absence of grazing or cutting. Also occurs on other suitable substrates in meadows, pastures, sand dunes, road sides, quarries, etc.

D Widespread in the south and east of England, generally scarce in north and west Wales, and almost absent as a native in Scotland.

S Native. Some increase since myxomatosis in the 1950s, and also as a result of reduced grazing over areas of farmland and downlands.

75a. Ligules long (to 4 mm), toothed; upper sheath not hairy; young blades slightly twisted .... ***Anisantha sterilis***
(*Bromus sterilis* • •)
Barren Brome

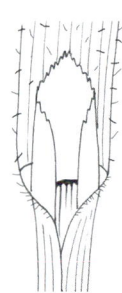

75b. Ligules shorter, not toothed 76

76a. Sheaths densely and very softly hairy; ligule up to 2.5 mm, hairy. Grasslands and waste places ..... ***Bromus hordaeceus***
(*Bromus mollis* •)
Soft Brome/Lop Grass

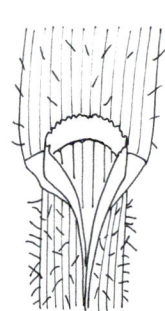

76b. Sheaths only loosely hairy 77

77a. Ligule shorter than wide, torn; blade 3-9 mm wide. Grasslands .... ***Bromus commutatus***
Meadow Brome

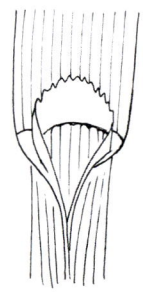

77b. Ligule longer than wide; blade narrower (2-5 mm wide). Grasslands .... ***Bromus racemosus***
Smooth Brome

Comment: This can be a difficult separation even in flower. In Stace 2019 these have been amalgamated into *B. racemosus*.

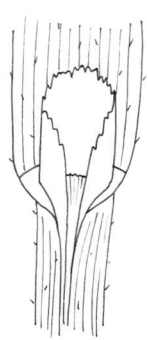

Anisantha sterilis

An annual with long, toothed ligule. Leaves soft, flaccid, floppy and softly hairy but sheath generally glabrous.

H A common weed of open habitats, though usually on well drained substrates. Common in hedgerows, roadsides, field margins, disturbed ground, etc.

D Generally lowland throughout England and Wales and east Scotland. Scarce in north and west Scotland and the mountains of Wales, north England and Scotland.

S Archaeophyte, stable.

Bromus hordaeceus

Winter annual. Generally the most hairy of the *Bromus* species being hairy throughout, with particularly dense hairs on the lower sheath. Very variable in size and habit.

H Open habitats on fertile, disturbed soils; tends to avoid wet or acidic substrates. Occurs frequently in meadows, pastures and also dunes, cliff top grasslands, waste ground, gardens, etc.

D Throughout, but scarce in central and northern Scotland.

S Native and stable.

Bromus commutatus

Sheaths with short hairs only. Hard to separate from *B. racemosus* and in Stace 4th edition they have been regrouped into a single species again.

H Damp meadows, lowland.

D Lowland. SE England extending north to Yorkshire. Scattered in the SW Scarce in Wales.

S Native. Declining due to habitat loss of wet meadows.

Bromus racemosus

Sheaths with short, straight, spreading hairs at 90 degrees to the surface.

H Lowland, unimproved damp meadows on periodically flooded alluvial soils.

D Similar to *B. commutatus*.

S Native, declining since the 1960s due to agricultural improvement of meadow habitat.

Blade folded in sheath (from Couplet 8) ▼

78a. Lower leaf-sheaths loosely to densely hairy 79

78b. Lower leaf-sheath hairless, or only minutely hairy 81

79a. Blades narrow (1-2 mm), folded or rolled; sheaths with very short hairs .... **Koeleria macrantha**
(*Koeleria cristata* •)
Crested Hair-grass

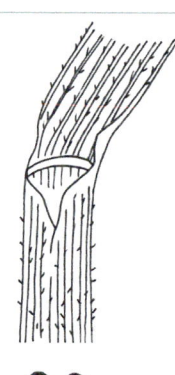

79b. Blades wider (2-6 mm); sheaths with spreading or deflexed long hairs 80

80a. Ligules long (to 8 mm), pointed; blades soft, 2-6 mm wide, glossy beneath; sheaths conspicuously hairy ....
.... **Avenula pubescens**
(*Helictotrichon pubescens* •)
Downy Oat-grass

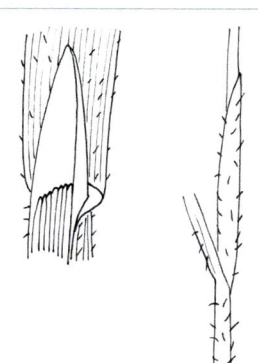

80b. Ligules shorter < 3mm, blunt; blades tough, narrow (2-3 mm wide), dull green with stiff bristles on their margins; sheaths sparsely hairy. Calcareous grassland .... **Bromopsis erecta**
(*Bromus erectus* • •)
Upright Brome

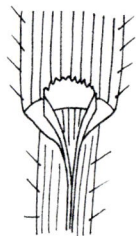

Koeleria macrantha

A slender, often stiff grass, with a densely hairy sheath – the hairs spreading outwards. Very variable in its growth form and leaf characteristics such that it also keys out in the 'needle leaf' section where leaves on non-flowering stems are often in-rolled and <1 mm wide. Leaves may be glaucous or green, glabrous or hairy, often with pronounced ribs.

🄗 Dry sandy areas near the sea, including sand dunes and cliff top grasslands. Also in open habitats on calcareous substrates or freely drained shallow soils of low fertility; in pastures and also on rocky outcrops, scree slopes and old mining spoil.

🄓 Generally lowland and coastal. Inland distribution is concentrated on the chalk and limestone of central England and Yorkshire.

🄢 Native and stable. Some recent decline due to habitat loss.

Avenula pubescens

A stiff, often tufted, grass. Initial impression is of a slightly hairy *Poa* due to the folded leaves and often blunt leaf tip. However, its hairiness separates it from *Poa*, and also from *Avenula pratensis*; which it is further distinguished from by its longer ligule – up to 8 mm, compared to only 2-5 mm in *Avenula pratensis*.

🄗 Wide range of grassland habitats on generally infertile, neutral to calcareous soils, with moderately good drainage. Mainly in meadows, pastures, dunes and cliff top grasslands; occasionally in woodland clearings and damp, marginal hill land.

🄓 Throughout but more frequent in lowland east England, west Wales and west Scotland.

🄢 Native and stable.

Bromopsis erecta

A densely tufted, winter green, perennial. Lower leaves tend to be narrow and inrolled but upper leaves, when flattened, have very distinct bristles along their margins which arise at c.45 degrees to the leaf edge giving a 'fish bone' appearance. This character helps distinguish it from *Brachypodium pinnatum* which grows in similar habitats and also has a hairy sheath and narrow, often inrolled, leaves.

🄗 Infertile, calcareous soils on chalk and limestone where it may become dominant in the absence of grazing or cutting. Also occurs on other suitable substrates in meadows, pastures, sand dunes, road sides, quarries, etc.

🄓 Widespread in the south and east of England, generally scarce in north and west Wales, and almost absent as a native in Scotland.

🄢 Native. Some increase since myxomatosis in the 1950s, and also as a result of reduced grazing over areas of farmland and downlands.

81a. Leaves usually with auricles; basal sheaths mostly pink or purple. Very common .... ***Lolium perenne***
Perennial Rye-grass

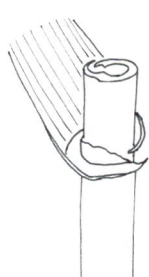

81b. Leaves without auricles 82

82a. Basal leaf-sheaths streaked with yellow in the lower part; blades 1-4 mm wide, glossy beneath, leaf often with a 'pinch point' in the upper part .... ***Cynosurus cristatus***
Crested Dog's-tail

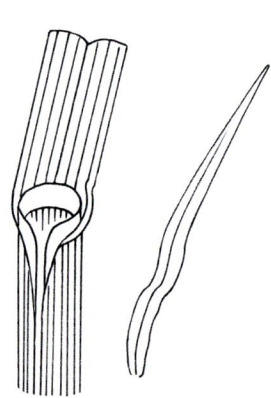

82b. Basal leaf-sheaths whitish, greenish or purplish 83

83a. Leaf-sheaths, on inner side, with cross veins and air cavities between the main nerves. Wetland places 84

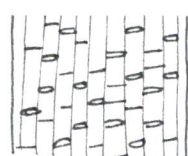

83b. Leaf-sheaths without cross veins or air cavities 88

Lolium perenne

The combination of folded, shiny green leaves, small auricles and purple sheath base is usually sufficient for quick identification of this common grassland species amongst the other grasses with which it commonly grows. Neutral grassland indicator.

🄷 Neutral but generally fertile soils but will also grow on soils of modest acidity or base-enrichment. Especially abundant in agriculturally improved grasslands, pastures, leys, amenity grasslands, etc. Also a natural component of many neutral grassland communities.

🄳 Throughout.

🅂 Native and stable.

Cynosurus cristatus

A rather anonymous grass, often overlooked when not in flower. The folded leaf expression may be missed for *Lolium perenne* but *Cynosurus* leaves are only shiny on the lower surface, they lack auricles and the sheath base tends to be nicotine yellow rather than the red of *Lolium*. Another useful character is the 'pinch' point about 2/3 up the leaf. Neutral grassland indicator.

🄷 Grasslands, especially pastures and hay meadows. Widespread but avoids extremes of soil water logging, drought and pH.

🄳 Throughout; generally lowland but up to 850 m in Westmoreland.

🅂 Native and stable.

84a. Leaf sheath open; ligule shorter than wide (2-8 mm), oval, often notched; leaves strongly hooded splitting at apex. Wetland places .... ***Catabrosa aquatica***
Whorl-grass

84b. Leaf sheath fused, though may split at the top; ligules various but not oval 85

85a. Rhizomatous with erect shoots. Ligule broad, not exceeding 6 mm, blunt with a fine tooth at its centre; blades wide, 7-20 mm; culms stout .... ***Glyceria maxima***
Reed Sweet-grass

85b. Stoloniferous with prostrate shoots. Ligules longer, 5-15 mm, blunt or pointed but lacking tooth; blades narrow; culms relatively slender 86

86a. Leaf sheath rough or slightly hairy near blade; leaves hairy at least below, yellowish green; ligule short (2-8 mm), bluntly rounded, whitish; shoots prostrate and root at nodes ....
.... ***Glyceria notata***
(*Glyceria plicata* • •)
Plicate Sweet-grass

86b. Leaf sheath smooth; leaves hairless but may be rough on margin; ligule pointed (5-15 mm) 87

Catabrosa aquatica

Stoloniferous grass, with stems often floating. The open leaf sheath distinguishes it from the species of *Glyceria* with which it shares the presence of cross veins and air chambers in the upper leaf sheaf. Leaves are often grey-green, generally short. The blunt leaf tip often tears to produce a small notch.

H Muddy habitats along pond margins, ditches and slow flowing streams.

D A lowland species more or less restricted to central England, East Anglia, and the coasts of west Scotland. Scarce in Wales and the SW Peninsula.

S Native, declining due to habitat loss, especially pond drainage/infilling and water course canalisation.

Glyceria maxima

A robust grass spreading by stout rhizomes, often forming extensive stands on pond margins and temporarily inundated grasslands. Distinguished from other species of *Glyceria*, not only by its size and upright growth form but also its very distinct ligule which is short and blunt but with a sharp central point.

H Wet habitats including pond and lake margins, ditches etc where it may form floating mats. Also in flooded grasslands. Formerly planted as a fodder crop due to its succulent and palatable foliage.

D Generally lowland. Scarce in the uplands of Wales, north England, SW England and Scotland.

S Native. Increased since the 1960s due to planting and natural expansion of planted populations.

Glyceria notata

Sheaths are slightly rough but it is the ligule shape that is probably the best character to distinguish this from *G. fluitans* and *G. declinata* as both those species have a pointed ligule. The pointed leaf tip further distinguishes it from *G. declinata*. *G. notata* readily hybridises with *G. fluitans* producing male sterile hybrids with hybrid vigour often forming luxuriant vegetative growth.

H Ditches, streams and muddy habitats, more tolerant of calcareous substrates than the other species.

D Throughout lowland England, Wales and Ireland. Scarce in Scotland except for the east coast and central Lowlands.

S Native. Some decline in south England due to drainage.

87a. Leaves green, tapering to pointed tip, ribbed above; sheaths tubular; ligule long (5-15 mm), pointed; shoots more or less erect .... ***Glyceria fluitans***
Floating Sweet-grass

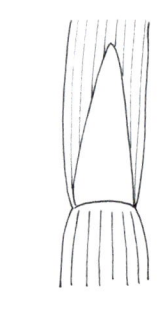

87b. Leaves greyish green, parallel sided, blunt or abruptly pointed at tip, not ribbed; sheaths keeled; ligule shorter (4-9 mm); cross veins weak; shoots usually curved .... ***Glyceria declinata***
Small Sweet-grass

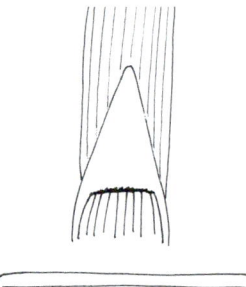

Leaf sheaths without cross veins or air cavities ▼

88a. Ligules barely distinguishable, but fringed with minute hairs; calcareous grassland. Restricted to northern England .... ***Sesleria caerulea***
(*Sesleria albicans*)
Blue Moor-grass

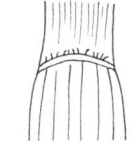

88b. Ligules apparent, if short then without a hairy fringe 89

Glyceria fluitans

Hairless sheaths and long pointed ligule separate this species from *G. notata* whilst the pointed, rather than blunt, leaf tips and green, rather than grey green colour helps separate it from *G. declinata*. Sometimes the parallel sidedness of the leaves in *G. declinata* can be easier to see than the blunt tip.

🄷 Marshes, swamps and pond margins. Also forms floating mats in shallow ditches and ponds. Also colonises grasslands subject to prolonged periods of inundation or moderate levels of disturbance.

🄳 Throughout.

🅂 Native and stable.

Glyceria declinata

Leaf shape and colour are the best vegetative characters for separation from *G. fluitans*, though this is not always easy. Observation of the parallel sidedness of the leaves of *G. declinata* is often easier to use than the blunt tip. The best diagnostic character is the lemma shape in flowering material.

🄷 Muddy open habitats in ditches, ponds, etc., and also bare areas in muddy fields and cattle trampled areas.

🄳 Less frequent than *G. fluitans*. More western in its distribution than *G. notata*; occuring throughout Wales, SW England and the south of Scotland, extending further north in more coastal and lowland situations.

🅂 Native. Under-recorded. Some declines in east England.

Sesleria caerulea

A distinctive grass with stiff, blue-green, short, parallel sided and hooded leaves (cf *Helictotrichon pratensis* but that species has an open sheath in contrast to the closed sheath of *Sesleria*). The ciliate ligule is the key character which distinguishes it from all the common species of *Poa*, otherwise only rare species of *Poa* are likely to be confused here.

🄷 Well drained limestone grassland, limestone pavement, screes and cliffs. Occasionally in woodland and coastal sandy areas in Ireland. When unmanaged may form dense stands of tough, unpalatable grass..

🄳 More or less restricted to areas of Carboniferous limestone in north England (Cumbria, Yorkshire) and Magnesian limestone in Durham and Northumberland. Also occurs on the Burren in Ireland and outliers on mica schist in Perthshire.

🅂 Native and stable.

89a. Plants with slender creeping rhizomes 90

89b. Plants without rhizomes; may be tufted or stoloniferous 92

90a. Rhizomes extensive; vegetative shoots appear solitary; sheaths usually with hairs at leaf junction. Rather local .... ***Poa humilis***
(*Poa subcaerulea* • •)
Spreading Meadow-grass

90b. Rhizomes less extensive; vegetative shoots appear clustered; sheaths glabrous at leaf junction 91

91a. Leaves mostly bright green; ligules usually <1 mm; blades soft, each several times longer than its accompanying sheath. Common .... ***Poa pratensis***
Smooth Meadow-grass

91b. Leaves mostly blue/ greyish green; ligules longer, up to 3 mm; blades stiff, each less than twice as long as its accompanying sheath or less. Dry habitats .... ***Poa compressa***
Flattened Meadow-grass

Poa humilis

Narrow, blue-green leaves, often with scabrid margins. Diagnostic characters are the hairs at the extreme base of the leaf blade where it joins the sheath, and the hairs on the back of the ligule. Only possibly confused with *P. angustifolia* (the third of the *Poa pratensis* group) but that species is green, rather than grey green and lacks the ciliate ligule and scabrid leaf margins of *P. humilis*.

H Grasslands on generally freely drained, often slightly calcareous, substrates. Including meadows, pastures, coastal grasslands and lower slopes in montane districts.

D Generally lowland, especially coastal habitats but extends into the marginal uplands.

S Native. Probably stable but unclear due to poor mapping and segregation within the *Poa pratensis* agg.

Poa pratensis

P. pratensis has the 'classic' appearance of a *Poa* with clearly folded leaves which are obviously hooded at their tips. Lacks hairs on the leaves and sheaths. Ligule is shorter than in *P. compressa* and is decurrent on the sheath margin.

H Wide habitat range in terms of soil base status and fertility but generally absent from the wetter grasslands.

D Throughout.

S Native and stable.

Poa compressa

A stiff, strongly rhizomatous, perennial grass lacking hairs. Leaves are grey-green, dull below and the sheath is often reddish at the base. Leaf blades are thin and pointed, lacking the obvious 'hood' of *P. pratensis*.

H Dry habitats on poor, thin or stony soils, often in man made or disturbed habitats including grassy banks, walls, etc.

D Lowland and scarce in the mountains of Scotland, Wales and SW England.

S Native but introduced in Scotland and Ireland making it difficult to assess changes in distribution.

92a. Plants densely tufted 93

92b. Plants not densely tufted but may be stoloniferous 95

93a. Ligule short (<5 mm); blades glaucous above, green below, rigid, rather narrow (1-5 mm) with blunt tip. Dry grassland .... ***Avenula pratensis***
(*Helictotrichon pratensis* •)
Meadow Oat-grass

93b. Ligules long (up to 15 mm); blades green, broader (up to 10 mm) with pointed tips 94

94a. Shoots broad, very flattened and softly succulent, basal sheath white, margins united. Leaves wide (up to 14 mm); ligule long, blunt (up to 12 mm); tussock forming. Grassland species, very common .... ***Dactylis glomerata***
Cock's-foot

94b. Leaf blade prominently ribbed and very rough above, stiff, narrow (up to 5 mm); ligule up to 15 mm, narrow, pointed; tussock forming. Grassland species .... ***Deschampsia cespitosa***
Tufted Hair-grass

Avenula pratensis

A tufted perennial with a superficial resemblance to *Poa pratensis*, with folded, blunt tipped leaves and although lacking rhizomes these are not always easy to see in *Poa*. However, *Avenula pratensis* has a glaucous leaf and distinct ligule. From all species of *Poa* the best character is the distinct tram lines on the upper leaf surface in *Avenula pratensis*. Although species of *Pucinellia* also have distinct ribs on their upper surface the leaves are much narrower and habitats are unlikely to overlap. Calcareous grassland indicator.

H Dry to moist, neutral to calcareous, soils. Common in meadows, pastures, dune and seacliff grasslands, open woodland and occasionally on less acidic heathlands. Resistant to grazing and mowing but not to enhanced fertility through fertilisation.

D Generally lowland and eastern, concentrated on areas of chalk and limestone but can also occur on lime-rich drift material. Scarce in Wales, SW Peninsula and north Scotland.

S Native and stable.

Dactylis glomerata

A distinctive grass with broad shoots and wide, soft, folded and pale green leaves. Ligule is long and blunt. Although forming large tussocks in unmanaged situations it can also occur in an intimate mixture with other grass species in pastures, meadows, etc. Neutral grassland indicator.

H Wide range of habitats, usually on well drained, neutral soils, including woodland ridges, meadows and pastures, coastal habitats and especially common on roadsides and waste places.

D Throughout. Some extreme northern and western locations may be relicts from former cultivation and seeding.

S Native and stable.

Deschampsia cespitosa

Compared to *Dactylis* this is a very tough and rough leaved grass – you risk cutting your fingers if you rub them backwards down the leaf blade, often a good way to identify its presence in a dense grass sward! The ligule is long and pointed and leaves deep, often slightly shiny, green. Ribs on leaf are prominent. This will often key out as a rolled leaf blade.

H Wet and poorly drained soils; marshy grassland, moorland and woodland.

D From sea level to 1,200 m in Scotland.

S Native and stable.

95a. Ligules short, 0.5 mm or less; blades narrow, 1-3 mm wide; loose tufts. Woods and shady places .... ***Poa nemoralis***
Wood Meadow-grass

95b. Ligules longer (up to 10 mm); blades wider, 1-6 mm wide 96

96a. Short annual, 3-30 cm tall; basal leaf-sheaths usually whitish, upper sheaths smooth; ligules 1-5 mm; blades transversely wrinkled when young, with blunt or hooded tips .... ***Poa annua***
Annual Meadow-grass

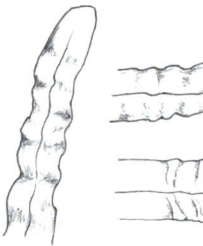

96b. Perennial, blade and growth form not as above 97

97a. Tall perennial, up to 100 cm; with leafy stolons. Ligules long, pointed (4-10 mm); leaves sharply pointed, green, up to 6 mm wide; basal leaf-sheaths usually reddish, upper sheaths often rough .... ***Poa trivialis***
Rough Meadow-grass

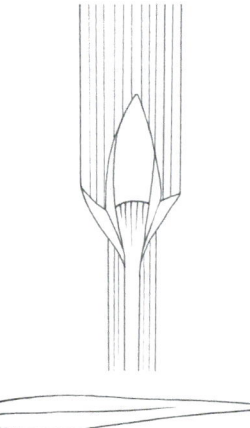

97b. Shorter perennial, with or without stolons. Ligule short, blunt (1-3 mm); leaf blades short (2-20 cm), narrow (1-4 mm), abruptly pointed or hooded at their tips, glaucous on both sides, ribbed. Saltmarshes (or salty roadsides) 98

Poa nemoralis

A very slender upright grass forming loose tufts; leaves are narrow, with pointed tips which contrasts it with most other species of *Poa*. Many surveyors are tempted, in woodland rides, to assign young *Poa trivialis* to *P. nemoralis* because in the shade the leaves of *P. trivialis* are narrow and the ligule is poorly developed, but *P. trivialis* is still more robust, and clearly distinguished by its creeping growth habit and the shiny lower surface to the leaf blade.

H Shady habitats; woodlands, hedgerows and occasionally on mountain ledges.

D Throughout but generally lowland. Scarce in north and west Scotland. Some introductions due to its ornamental inflorescence.

S Native, probably stable. Introduced in Ireland.

Poa annua

A small annual, often pale whitish green in colour. The distinguishing character is the wrinkling of the leaves which is only otherwise found in *Poa infirma* (a rare annual of coastal habitats in the south of England). May sometimes act as a perennial.

H Disturbed and artificial habitats, especially grasslands, amenity areas, roadsides, paths, etc. Also areas of maritime and coastal grassland where it often functions as a perennial.

D Throughout.

S Native and stable.

Poa trivialis

An obviously creeping grass when growing in damp, open or shady conditions, though stolons are often obscure when growing in a dense grassy sward e.g. in meadows and pastures. The sheath and shiny lower side of the leaf blade may give the impression of a young *Lolium perenne* but *P. trivialis* lacks auricles and the intensely purple basal sheath of *Lolium*. Compared to *P. nemoralis* the leaves are broader and spread is via stolons though surveyors often try and classify young, shade grown, *P. trivialis* as *P. nemoralis*.

H Wide range of habitats but tending to be more abundant on damp soils in woodlands, meadows, pastures and marshy grassland.

D Throughout except some high mountain areas of Scotland. Introduced in commercial seed mixes and a common wool alien.

S Native and stable.

98a. Loosely spreading on long creeping stolons that root at the nodes forming a compact turf, sometimes tuft forming. Salt marshes .... ***Puccinellia maritima***
Common Saltmarsh-grass

98b. Lacks creeping stolons; shoots erect or prostrate. Saline habitats but often found away from salt marshes .... ***Puccinellia distans***
Reflexed Saltmarsh-grass

Leaf blades bristle like, narrow (from Couplet 1)

99a. Shoots in small tufts or solitary without old sheaths at base; annuals 100

99b. Tufted or mat forming, with remains of old sheaths and culms at the base; perennials 104

100a. Ligules short (< 1 mm); leaves ribbed above 101

100b. Ligules longer 1-5 mm; leaves not ribbed above. Sandy places 103

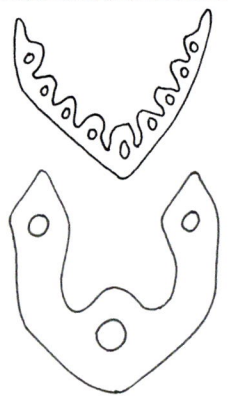

Puccinellia maritima

An obviously stoloniferous grass with extensive runners that freely root at their nodes. This growth form distinguishes it form *P. distans* and the other less common *P. fasciculata* and *P. rupestris* (species that are both more or less restricted to the area south and east of a line between the Wash-Severn estuary).

H Lowland, saltmarshes. Generally on the lower and middle marsh where it forms extensive lawns. Locally on saline pans and along ditches above the high water mark.

D Coastal, throughout.

S Native and stable. Has recently been recorded colonising salt treated roadsides in England.

Puccinellia distans

Lacks stolons and tends to form small, discrete, clumps, often away from salt marsh habitats. Further distinguished from *P. maritima* (and *P. rupestris*) by the more obvious ribs on the upper leaf surface and from *P. fasciculata* by its longer ligule.

H Lowland on poorly drained, compacted soils, on upper saltmarshes and coastal rocky areas, and as a colonist of road sides.

D Coastal but less widespread than *P. maritima* in that habitat. Scarce on the west coast of north England and Scotland. Also extensive inland along roadsides.

S Native. Stable in coastal habitats.

As an alien along inland roadsides where it has expanded since the 1970s.

101a. Leaves minutely hairy above, blade pointed, flat to inrolled; ligule entire with swelling at top of sheath; sheath open. Arable and waste places .... ***Vulpia*** spp.

101b. Leaves not hairy, blade strongly u-shaped; ligule slightly jagged; sheath closed. Dry sandy places 102

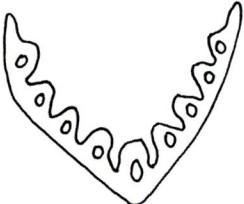

102a. Leaf blade green or purple, finely ribbed, narrow (0.5-2 mm), pointed, incurved and sickle shaped. [Panicle one sided and branched at base.] Dry places .... ***Catapodium rigidum***
(*Desmazeria rigida* •)
Fern-grass

102b. Leaf blade dark green, fleshier with pronounced ribs above, wider (1-3 mm), blunt tipped. [Panicle narrower and unbranched.] Dry coastal sand, walls and rocky areas ....
.... ***Catapodium marinum***
(*Desmazeria marina* •)
Sea Fern-grass

Vulpia spp.

Three relatively common species; *V. bromoides*, *V. myuros* and *V. fasiculata*. All are minutely hairy on the upper leaf surface which distinguishes them from *Aira* sp, *Catapodium* sp and *Mibora*. Individually they are not easy to separate vegetatively but, being annuals, flowers are usually present!

V. bromoides Squirrel-tail Fescue (illustrated)

Leaves with c.9 ribs, sheaths not inflated. Ligule entire.

H Well drained soils in open situations; coastal grasslands, heaths, waste ground, quarries, etc.

D Throughout England and Wales. In Scotland tends to be restricted to coastal and lowland areas.

S Native and stable.

V. myuros Rat's-tail Fescue

Leaves with 9-11 ribs, sheaths not inflated. Ligule toothed.

H Waste places, including railways, pavements, and rarely as a weed of cultivation.

D Lowland. England and Wales, scarce in Scotland.

S Archaeophyte. Expanded since the 1960s largely along the railway network.

V. fasiculata (*V. membranacea* •) Dune Fescue

Leaves with c.17 ribs; sheaths inflated. Ligule very short.

H Fixed dunes and sandy shingle.

D Coastal; England and Wales. Absent from Scotland and N. Ireland.

S Native and stable. Some local increases in frequency following decline in rabbit populations.

Catapodium rigidum

Another winter annual. Leaves are narrower and pointed compared to *C. marinum* but a good diagnostic character is the incurved leaf blade which curves like a sythe. The panicle is more obviously branched.

H Infertile, bare substrates in open, coastal areas including dry banks, cliff tops, sand dunes, and on walls, pavements, etc. Often occurs with other dune annual grasses.

D Lowland, mainly coastal, especially in the south and west of England, Wales and Ireland. Scarce in NE England and NE Scotland. Inland along salt treated roadsides.

S Native and stable. Localised expansion in the SE of England along roadsides in the M4 corridor.

Catapodium marinum

Overwintering as small plants. Generally more robust than *C. rigidum*; the blunt leaf tip gives the appearance of a small *Poa* but the leaves here are rolled not folded. In flower the narrower, unbranched panicle is a good character.

H Dry, open habitats, often on calcareous substrates.

D Lowland, throughout southern England, coastal Wales and Northern Ireland. Scarce in Scotland where it reaches its world northern limit at the Forth of Tay.

S Native and stable.

103a. Leaf-sheaths rough; blades greyish green; ligule toothed, up to 5 mm; plants up to 40 cm tall ...... ***Aira caryophyllea***
Silver Hair-grass

103b. Leaf-sheaths smooth; blades green or reddish; ligule not toothed, up to 3 mm; more slender plant rarely exceeding 20 cm tall .... ***Aira praecox***
Early Hair-grass

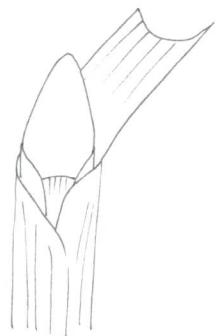

104a. Roots yellow, tough, cord-like; lower sheaths shiny, hard, white; leaf blades short-pointed in tight tussocks, rough, ridged, often pointing outwards at base. Moors and heaths .... ***Nardus stricta***
Mat-grass

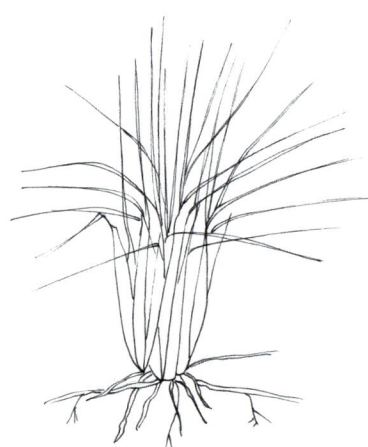

104b. Roots fine; basal sheaths not hard and shiny 105

Aira caryophyllea

Distinguished from the very similar *A. praecox* by its longer, toothed ligule and rough sheath. When in flower the spreading inflorescence is diagnostic. Despite the name, both *A. praecox* and *A. caryophyllea* tend to have silvery leaves when young.

H Well drained, sandy habitats including sand dunes, cliff tops, summer parched grasslands especially round rocky outcrops, anthills, walls, etc.

D Throughout but tends to be coastal and lowland in Scotland.

S Native, Declined since the 1950s, especially in SE England.

Aira praecox

Subtly different from *A. caryophyllea* in its entire ligule and smooth sheath. Often recognised by small, shiny patches, of clump forming leaves in the spring.

H Generally thin, well drained sandy and rocky habitats, especially acidic grasslands, cliff tops, sand dunes and heaths.

D Throughout but scarce in a band of lowland England SE of the Wash-Severn line.

S Native. Some decline in SE England but less so than for *A. caryophyllea*.

Nardus stricta

A very tough, tussock forming grass with leaves that are rough to touch, especially if running your fingers down the leaf (a good way of finding it in a closed sward). Basal sheaths are white and very shiny, and extremely tough and the leaves often come out almost horizontally from the sheath. Hard to pull up but the thick, yellow (cord-like) roots are diagnostic.

H Poorly drained, generally acidic, peaty, podzolic or poor brown earth soils of low fertility. It is unpalatable and this enables it to dominate areas of heath and grassland that are over-grazed.

D Throughout Scotland, Wales and SW Peninsula of England. In south and central England restricted to areas of suitable soils on heaths and acidic grasslands

S Native, stable, but with some local declines in lowland England.

105a. Ligules long, up to 3 cm, pointed; blades greyish or whitish, sharply pointed, tightly inrolled when young, with minute hairs on upper ribs; extensive rhizomes. Sand dunes .... ***Ammophila arenaria***
Marram Grass

105b. Ligules shorter, up to 4 mm, or absent 106

106a. Ligule generally very short, <0.5 mm or absent 107

106b. Ligule 0.5-4 mm long 109

107a. Leaf sheath tubular, rounded, basal sheaths purple, shortly hairy or hairless; blades above inrolled, 0.5-1 mm wide opening to 2 mm wide, minutely hairy; dark green or grey green; ligule obscure. Grasslands, common, especially in coastal habitats .... ***Festuca rubra*** agg.
Red Fescue

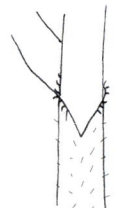

107b. Leaf sheath with free margins, hairless, rounded or folded 108

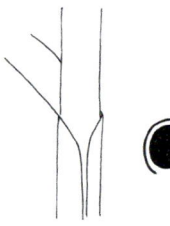

Ammophila arenaria

A large grass forming dense patches through extensive rhizome growth. The extremely long ligules, up to 3 cm, are a good diagnostic feature. Leaves grey-green up to 90 cm long. Sheaths overlapping and leaf tips very sharp. Leaves tightly inrolled in the sheath but often open out when mature, but, at 6 mm width, are much narrower than those of *Leymus arenarius* with which it often grows on the most mobile dune systems.

H Mobile sand dunes.

D Throughout lowland coastal regions.

S Native and stable. Attempts at inland colonisation on golf courses have generally failed.

Festuca rubra agg.

The tubular leaf sheath is diagnostic, and although it often splits near the blade on older leaves there is no overlapping of 'free margins' so apparent on *F. ovina*. Generally sparsely to densely shortly hairy on the upper leaf surfaces and sheaths. A rather variable character but in general density of hairs are greater than on *F. ovina*. Often forms dense carpets and tussocks when unmanaged.

H Grasslands on a wide range of substrates, generally avoiding soils which are subject to prolonged water logging. Common in meadows, pastures, coastal habitats including dunes, cliff tops and saltmarshes. Also roadsides and waste places

D Throughout.

S Native and stable.

108a. Densely tufted; without rhizomes or stolons; sheath rounded; leaf blades bristle-like, tightly inrolled, very narrow 0.3-0.6 mm; ligule obscure or absent. Common plant of dry, infertile habitats .... ***Festuca ovina*** agg.
Sheep's-fescue

108b. Plants with slender whitish, or brown scaly rhizomes; basal sheath keeled, shoots compressed; blades folded along mid-rib, up to 1-2 mm wide; upper ligules up to 1 mm long. Dry grassland .... ***Poa angustifolia***
Narrow-leaved Meadow-grass

109a. Basal leaf-sheaths softly and shortly hairy; leaves deeply ribbed on upper surface. Dry grassland ....
.... ***Koeleria macrantha***
(*Koeleria cristata* •)
Crested hair-grass

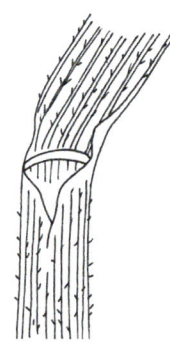

109b. Basal leaf-sheaths hairless 110

Festuca ovina agg.

Seperated from *F. rubra* by the open leaf sheath with free margins that wrap around the stem. Not always easy to see! Leaves may also be sparsely hairy, but generally less so than in *F. rubra*. Lacking rhizomes *F. ovina* tends to occur as discrete tufts, especially in more open grasslands and heathlands. Smooth sheaths and leaves help distinguish it from young plants of *Nardus stricta* with which it often occurs.

H Generally freely drained, infertile and often acidic, substrates. In a range of habitats including upland heaths and grasslands, calcareous grasslands, lowland coastal habitats. More drought resistant than *F. rubra* it tends not to occur on the wetter heaths and mires. Resistant to heavy grazing.

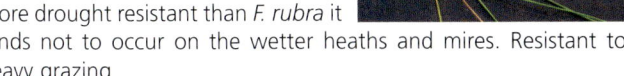

D Throughout.

S Native and stable.

Poa angustifolia

The 3rd of the sub-types of *Poa pratensis*. Superficially resembles *Festuca rubra* in its long, narrow leaves but in *P. angustifolia* the leaves are folded rather than rolled and the sheath is open not closed. From *Poa pratensis* by its narrower and longer leaves and longer ligule, and greater number of veins on the upper leaf surface.

H Infertile, freely drained, soils. Mainly grasslands, wall tops, railway lines, etc.

D Lowland. Concentrated in England south and east of the Tees-Exe line. Scattered in Wales and Scotland.

S Native. Possibly declining in its semi-natural habitats.

Koeleria macrantha

A slender, often stiff grass, with a densely hairy sheath – the hairs spreading outwards in contrast to the down-pointing hairs on *Trisetum*. Very variable in its growth form and leaf characteristics such that it also keys out at couplet 64 with other grasses with an open, hairy leaf sheath. Leaves may be glaucous or green, glabrous or hairy, deeply ridged on upper surface.

H Dry sandy areas near the sea, including sand dunes and cliff top grasslands. Also in open habitats on calcareous substrates or freely drained shallow soils of low fertility; in pastures and also on rocky outcrops, scree slopes and old mining spoil.

D Generally lowland and coastal. Inland distribution is concentrated on the chalk and limestone of central England and Yorkshire.

S Native and stable. Some recent decline due to habitat loss.

110a. Leaf blade very narrow, <0.3 mm, channelled, rough below, often pale; densely tufted. Heaths and moors in the south and southwest of England and Wales ...... **Agrostis curtisii**
(*Agrostis setacea* •)
Bristle bent

110b. Blade wider, 0.3-1 mm, smooth; generally darker colour 111

111a. Tufted without rhizomes or stolons 112

111b. With rhizomes or stolons, tufted or mat forming; ligule up to 4.0 mm 113

Agrostis curtisii

Stands out for its extremely slender growth form. The very narrow, pale green, leaves form discrete tufts, often creating dense patches in open heathland and grasslands. From *Festuca tenuifolia*, which grows in similar habitats, by its long and pointed ligule. From *Avenella flexuosa* by its stiffer growth form and dull, pale green leaves.

H Infertile, sandy and peaty substrates which are generally freely draining. A colonist of bare habitats it will expand rapidly into wet heathland following disturbance (especially burning) and woodland clearings.

D Generally lowland. South and west England from West Sussex and Surrey to Cornwall and also south Wales.

S Native, some decline due to habitat loss.

112a. Ligule short 0.5 (-3.0) mm, blunt; blade soft, bright green, shiny and silky, 0.5-0.8 mm wide, slightly angled in cross section. Heaths and moors .... ***Avenella flexuosa***
(*Deschampsia flexuosa* • • •)
Wavy Hair-grass

112b. Ligule longer (1-3 mm); shoots erect or prostrate; leaves stiff, with blunt or hooded tips, wider (1-4 mm), often glaucous. Saltmarshes and saline habitats inland (along road sides) .... ***Puccinellia distans***
Reflexed Saltmarsh-grass

113a. Plants with extensive rhizomes, lacking stolons; leaves stiff, erect, flat; ligule pointed, ragged. Dry acid grassland and heath .... ***Agrostis vinealis***
(*Agrostis canina montana* •)
Brown Bent

113b. Plants without rhizomes but with slender stolons, forming turfs of soft leaves which root at the nodes 114

Avenella flexuosa

Distinct from the other narrow leaved grasses with which it commonly grows (*F. ovina*, *Nardus stricta* and *Agrostis vinealis*) by its usually very shiny (almost silky) leaves which are a deep green colour, and generally much less stiff than the other associates. The presence of a short ligule separates it from *Nardus* and *F. ovina* which lack ligules and from *A. vinealis* that has a long ligule.

H Often forms discrete clumps in open grasslands and woodlands. Generally acidic, nutrient poor, soils with relatively good drainage, on heaths, acidic grasslands and woodlands. Will colonise the dry tops of tussocks of *Molina* and *Eriophorum vaginatum* in moorland situations; *F. ovina* in contrast is rarely found in such situations.

D Throughout but scarce on the east coast from South Yorkshire to Essex and the central lowlands of England due to lack of suitable habitat.

S Native, stable but with localised losses due to destruction of lowland heathlands.

Puccinellia distans

Lacks stolons and tends to form small, discrete, clumps, often away from salt marsh habitats. Further distinguished from *P. maritima* (and *P. rupestris*) by the more obvious ribs on the upper leaf surface and from *P. fasciculata* by its longer ligule.

H Lowland on poorly drained, compacted soils, on upper saltmarshes and coastal rocky areas, and as a colonist of road sides.

D Coastal but less widespread than *P. maritima* in that habitat. Scarce on the west coast of North England and Scotland. Also extensive inland along roadsides.

S Native. Stable in coastal habitats. As an alien along inland roadsides where it has expanded since the 1970s.

Agrostis vinealis

Rhizomes often obscure and shoots may appear clumped in dense grass turf. Clearer to see in heathland habitats where it tends to grow within subshrub patches in contrast to *A. capillaris* which forms dense grassy turfs. Leaves are narrow, often needle-like. Ligule is longer than wide and pointed.

H Sandy or peaty soils of heaths, grasslands, open woodlands.

D Widespread. Accurate distribution uncertain since often poorly separated from *A. canina* when only sub-species status.

S Native.

114a. Ligule pointed, 2-4 mm; leaf blades flat or rolled, finely pointed at tip. Damp grasslands and open wet habitats .... ***Agrostis canina***
(*Agrostis canina canina* •)
Velvet Bent

114b. Ligule blunt, <3 mm; leaves parallel sided, rolled or folded and hooded at tip. Saltmarshes .... ***Puccinellia maritima***
Common Saltmarsh-grass

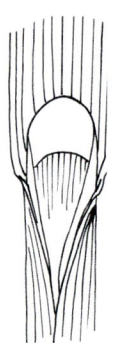

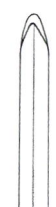

Agrostis canina

One of a trio of stoloniferous grasses of damp habitats. The leaf blade is narrower than *Agrostis stolonifera* and *Alopecurus geniculatus* and generally a brighter green colour. Forms 'soft' turf of compactly growing plants lacking the pink sheath of *A. stolonifera* and the waxy white of the *Alopecurus*. The ligule is usually pointed.

H Generally infertile habitats; permanent grasslands, heaths, mires and flushes on acidic peaty substrates, also open habitats on mineral soils which are permanently damp including tracks, ditches, water margins, etc.

D Throughout but scarce in the Midlands, NE England and the Highlands.

S Native. Probably stable but many earlier records for the aggregate make precise determination of change difficult.

Puccinellia maritima

An obviously stoloniferous grass with extensive runners that freely root at their nodes. This growth form distinguishes it form *P. distans* and the other less common *P. fasciculata* and *P. rupestris* (species that are both more or less restricted to the area south and east of a line between the Wash-Severn estuary). Distinguished from *Agrostis stolonifera*, which also grows on upper saltmarsh, by narrower, folded, blunt tip leaves and short, blunt ligule.

H Lowland, saltmarshes. Generally on the lower and middle marsh where it forms extensive lawns. Locally on saline pans and along ditches above the high water mark.

D Coastal, throughout.

S Native and stable. Has recently been recorded colonising salt treated roadsides in England.

Glossary of terms

Acute: Sharply pointed, refers to ligules and leaf tips ....

Air chamber: Transparent cross veins best observed on the inner surface of the lower leaf sheaths, just below the blade ....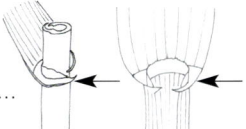

Air cavities: As above.

Annual: Plant that completes its life cycle in one year. They tend not to accumulate 'dead' leaf material at their base. Many germinate in the autumn and flower early in the spring referred to as 'winter annuals' ....

Auricles: Claw-like outgrowths where the leaf blade joins the leaf sheath ....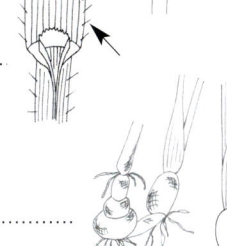

Blade: That part of the leaf above the sheath ....

Bristles: Stiff hairs often on the margin of the leaf cf. *Bromopsis erecta* ......

Bulbous: Swelling at the base of the stem. May be round (*Pheum*) or pear-shaped (*Arrhenatherum*) ....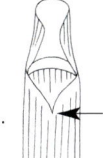

Ciliate: Fringed with hairs.

Closed sheath: When the margins of the leaf are fused together to form a tube around the stem. Sometimes referred to as a tubular sheath or fused ....

Compressed: Flattened, extremely folded ....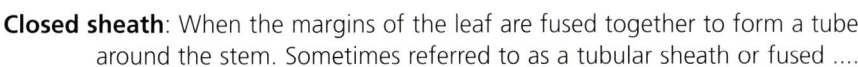

Cross veins: Short horizontal veins between the main longitudinal veins ....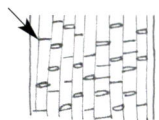

Culm: Flowering stem ....

Folded: Leaf blade folded lengthwise along the mid-rib with the upper surface 'within' ....

Free margins (leaf sheath): Where the leaf margins are wrapped around the stem, such that the edges of the leaf are clearly visible running down the stem (cf open sheath) ....

Fused (leaf sheath): When the margins of the leaf are fused together to form a tube around the stem. Sometimes referred to as a closed sheath ....

Glabrous: Without hairs.

Glaucous: Blue/green colour ....

Grooved: When the veins are very prominently raised ....

Hooded (leaf tip): Usually in leaves that are parallel sided and have a rounded leaf tip ....

Inflated (sheath): Swollen ....

Innovation: Vegetative shoots (c.f. Stace, Cope and Grey) ....

Inrolled (leaf): Leaf wrapped around itself with the upper face of the blade innermost ....

Leaf: The combined blade and sheath ....

Leaf blade: That part of the leaf above the sheath, sometimes referred to as the lamina ....

Leaf sheath: The extension of the blade downwards which surrounds the stem ....

Ligule: Membranous outgrowth arising on the inside of the leaf blade where it joins the leaf sheath ....

Lower sheath: Leaf sheaths associated with leaves near the base of the plant ....

Nerve: Slender veins or ribs on the leaf blade ....

Nodes: Points on the stem from which a leaf or branch arises ....

Obtuse: Blunt. Can refer to the leaf tip or the ligule ....

Open (leaf sheath): Where the leaf margins are wrapped around the stem, the edges of the leaf are clearly visible running down the stem ....

Perennial: A plant that lives for more than 1 year, and usually accumulates dead leaves around the base of the plant ....

Rhizome: Creeping stems spreading horizontally **below** ground surface. New shoots emerge from nodes along this ....

Ribbed: Where there are very prominent veins on the leaves ....

Scabrid: Covered in minute prickles or bristle like hairs ....

Sheath: The lower part of the leaf surrounding the stem ....

Smooth: Usually indicates a lack of hairs or of rough projections.

Stem: Main axis of the plant bearing leaves or flowering heads (culms) ....

Stolon: Creeping stems spreading horizontally **above** the ground surface that can root at the nodes and producing new vegetative shoots and culms ....

Tapering: Narrowing gradually ....

Tiller: Vegetative shoots ....

Tram lines: Parallel sided grooves between prominent nerves ....

Tubular (leaf sheath): When the margins of the leaf are fused together to form a tube around the stem. Sometimes referred to as a closed sheath ....

Tuft: Loose, compact or dense cluster of vegetative shoots ....

Tussock: Very dense cluster of vegetative shoots ....

Upper sheaths: Leaf sheaths associated with leaves near the top of the plant ....

Veins: Venation on grasses is parallel with the leaf margin. Often obscure, sometimes they are very prominent producing a grooved appearance. They may also have hairs on them ....

Youngest leaf blade: The upper most leaf emerging from the 'top' of the plant ....

Further reading

Clapham, A.R., Tutin, T.G. and Moore, D.M. (1987). *Flora of the British Isles*. 3rd edition. Cambridge University Press, Cambridge

Clapham, A.R., Tutin, T.G. and Warburg, E.F. (1995). *Excursion Flora of the British Isles*. Cambridge University Press, Cambridge.

Cope, T. and Grey, A. (2009). *Grasses of the British Isles*. B.S.B.I. Handbook No. 13. Botanical Society of the British Isles, London.

Fitzpatrick, U., Weekes, L. and Wright, M. (2016). *Identification guide to Ireland's grasses*. National Biodiversity Data Centre, Waterford.

Hubbard, J.C.E. (1984). *Grasses. A guide to their structure, identification, uses and distribution in the British Isles*. 3rd edition. Penguin Books, London.

JNCC (1990). *Handbook for Phase I Habitat Survey – a technique for environmental audit*. England Field Unit, Nature Conservancy Council.

Poland, J. and Clement, E. (2009). *The Vegetative Key to the British Flora*. John Poland, Southampton (in association with the Botanical Society of the British Isles).

Poland, J. and Clement, E. (2019). *The Vegetative Key to the British Flora*. 2nd edition. John Poland, Southampton (in association with the Botanical Society of the British Isles).

Preston, C.D., Pearman, D.A. and Dines, T.D. (2002). *New Atlas of the British and Irish Flora. An atlas of the vascular plants of Britain, Ireland, the Isle of Man and the Channel Islands*. Oxford University Press, Oxford.

Rodwell, J.S. ed. (1991 *et seq.*). *British Plant Communities*. Cambridge University Press, Cambridge.

Rose, F. (1989). *Colour identification guide to the grasses, sedges, rushes and ferns of the British Isles and north-western Europe*. Viking (part of Penguin Books, London).

Stace, C. (2010). *New Flora of the British Isles*. 3rd edition. Cambridge University Press, Cambridge.

Stace, C. (2019). *New Flora of the British Isles*. 4th edition. Cambridge University Press, Cambridge.

Index to species accounts

Agropyron caninum 44, 58
Agropyron junciforme 28
Agropyron pungens 18
Agropyron repens 18
Agrostis canina 44, 45, 94, 95
Agrostis canina canina 44, 94
Agrostis canina montana 34, 38, 92
Agrostis capillaris 36, 37, 42, 43
Agrostis curtisii 90, 91
Agrostis gigantea 36, 37, 40, 41
Agrostis setacea 90
Agrostis stolonifera 46, 47
Agrostis tenuis 36, 42
Agrostis vinealis 34, 35, 38, 39, 92, 93
Aira caryophyllea 84, 85
Aira praecox 84, 85
albicans, *Sesleria* 72
Alopecurus myosuroides 50, 51
Alopecurus geniculatus 46, 47
Alopecurus pratensis 44, 45, 60, 61
Ammophila arenaria 28, 29, 86, 87
anglica, *Spartina* 16, 17
angustifolia, *Poa* 88, 89
Anisantha sterilis 64, 65
annua, *Poa* 78
Annual Meadow-grass 78
Anthoxanthum odoratum 28, 29
Apera spica-venti 50, 51
aquatica, *Catabrosa* 70, 71
arenaria, *Ammophila* 28, 29, 86, 87
arenarium, *Phleum* 48, 49
arenarius, *Elymus* 18
arenarius, *Leymus* 18, 19
Arrhenatherum elatius 32, 33, 50, 51
Arrhenatherum elatius var *bulbosus* 30, 31
arundinacea, *Festuca* 24
arundinacea, *Phalaris* 34, 35
arundinaceus, *Schedonorus* 24, 25
athericus, *Elymus* 18, 19
australis, *Phragmites* 12, 13
Avena fatua 56, 57
Avenella flexuosa 92, 93
Avenula pratensis 76, 77
Avenula pubescens 66, 67

Barley, Meadow 22
Barley, Sea 26
Barley, Wall 22, 26
Barley, Wood 20
Barren Brome 64
Bearded Couch 44, 58
Bent, Black 36, 40
Bent, Bristle 90
Bent, Brown 34, 38, 92
Bent, Common 36, 42
Bent, Creeping 46
Bent, Velvet 44, 94
bertolonii, *Phleum pratense* 30, 31
Black Bent 36, 40
Black-grass 50
Blue Moor Grass 72
Brachypodium pinnatum 36, 37, 42, 43, 54, 55
Brachypodium sylvaticum 58, 59
Bristle Bent 90
Briza media 60, 61
Brome, Barren 64
Brome, False 58
Brome, Hairy 20
Brome, Meadow 64
Brome, Rye 60
Brome, Smooth 64
Brome, Soft 64
Brome, Upright 62, 66
bromoides, *Vulpia* 83
Bromopsis erecta 62, 63, 66, 67
Bromopsis ramosa 20, 21
Bromus commutatus 64, 65
Bromus erectus 62, 66
Bromus hordaeceus 64, 65
Bromus mollis 64
Bromus racemosus 64, 65
Bromus ramosus 20
Bromus secalinus 60, 61
Bromus sterilis 64
Brown Bent 34, 38, 92
bulbosus, *Arrhenatherum elatius* var 30, 31

caerulea, *Sesleria* 72, 73
caerulea, *Molinia* 14, 15
Calamagrostis canescens 52, 53

Calamagrostis epigejos 34, 35
Canary-grass, Reed 34
canescens, Calamagrostis 52, 53
canina canina, Agrostis 44, 94
canina montana, Agrostis 34, 38, 92
canina, Agrostis 44, 45, 94, 95
caninum, Agropyron 44, 58
caninus, Elymus 22, 23, 44, 45, 58, 59
capillaris, Agrostis 36, 37, 42, 43
caryophyllea, Aira 84, 85
Cat's-tail, Smaller 30
Catabrosa aquatica 70, 71
Catapodium marinum 48, 82, 83
Catapodium rigidum 48, 49, 82, 83
cespitosa, Deschampsia 38, 39, 76, 77
Cock's-foot 76
Common Bent 36, 42
Common Cord-grass 16
Common Reed 12
Common Saltmarsh-grass 80, 94
commutatus, Bromus 64, 65
compressa Poa 74
Cord-grass, Common 16
Cord-grass, Small 14
Cord-grass, Townsend's 16
Couch Grass 18
Couch, Bearded 44, 58
Couch, Sand 28
Couch, Sea 18
Creeping Bent 46
Creeping Soft-grass 34, 54, 62
Crested Dog's-tail 40, 52, 60, 68
Crested Hair-grass 56, 66, 88
cristata, Koeleria 56, 66, 88
cristatus, Cynosurus 40, 41, 52, 53, 60, 61, 68, 69
curtisii, Agrostis 90, 91
Cynosurus cristatus 40, 41, 52, 53, 60, 61, 68, 69

Dactylis glomerata 76, 77
Danthonia decumbens 14, 15
declinata, Glyceria 72, 73
decumbens, Danthonia 14, 15
decumbens, Sieglinglia 14
Deschampsia cespitosa 38, 39, 76, 77
Deschampsia flexuosa 92
Desmazeria marina 48, 49, 82

Desmazeria rigida 48, 82
distans, Puccinellia s 80, 81, 92, 93
Dog's-tail, Crested 40, 52, 60, 68
Downy Oat-grass 66

Early Hair-grass 84
effusum, Milium 40, 41
elatius var *bulbosus, Arrhenatherum* 30, 31
elatius, Arrhenatherum 32, 33, 50, 51
Elymus arenarius 18
Elymus athericus 18, 19
Elymus caninus 22, 23, 44, 45, 58, 59
Elymus farctus 28
Elymus junceiformis 28, 29
Elymus pycnanthus 18
Elymus repens 18, 19
Elytrigia juncea 28
Elytrigia repens 18
epigejos, Calamagrostis 34, 35
erecta, Bromopsis 62, 63, 66, 67
erectus, Bromus 62, 66
europaeus, Hordelymus 20, 21

False Brome 58
False Oat-grass 30, 32, 50
False-brome, Heath 36, 42, 54
farctus, Elymus 28
fasiculata, Vulpia 83
fatua, Avena 56, 57
Fern-grass 48, 49, 82
Fern-grass, Sea 48, 82
Fescue, Giant 24
Fescue, Meadow 24
Fescue, Sheep's 88
Fescue, Tall 24
Fescue, Red 86
Festuca arundinacea 24
Festuca gigantea 24
Festuca ovina agg. 88, 89
Festuca pratensis 24, 25
Festuca rubra agg. 86, 87
Flattened Meadow-grass 74
flavescens, Trisetum 40, 41, 56, 57
flexuosa, Avenella 92, 93
flexuosa, Deschampsia 92
Floating Sweet-grass 72

fluitans, *Glyceria* 72, 73
Foxtail, Marsh 46
Foxtail, Meadow 44, 60

geniculatus, *Alopecurus* 46, 47
Giant Fescue 24
gigantea, *Agrostis* 36, 37, 40, 41
gigantea, *Festuca* 24
giganteus, *Schedonorus* 24, 25
glomerata, *Dactylis* 76, 77
Glyceria declinata 72, 73
Glyceria fluitans 72, 73
Glyceria maxima 70, 71
Glyceria notata 70, 71
Glyceria plicata 70
Grass, Black 50
Grass, Couch 18
Grass, Heath 14
Grass, Lop 64
Grass, Lyme 18
Grass, Marram 28, 86

Hair-grass, Crested 56, 66, 88
Hair-grass, Early 84
Hair-grass, Silver 84
Hair-grass, Tufted 38, 76
Hair-grass, Wavy 92
Hairy-brome 20
Hairy-brome, Lesser 22
Hard-grass 48
Heath False-brome 36, 42, 54
Heath Grass 14
Helictotrichon pubescens 66
Helictotrichon pratensis 76
Holcus lanatus 52, 53, 62, 63
Holcus mollis 34, 35, 54, 55, 62, 63
hordaeceus, *Bromus* 64, 65
Hordelymus europaeus 20, 21
Hordeum marinum 26, 27
Hordeum murinum 22, 23, 26, 27
Hordeum secalinum 22, 23
humilis, *Poa* 74, 75

Italian Rye-grass 26

juncea, *Elytrigia* 28
junceiformis, *Elymus* 28, 29
junciforme, *Agropyron* 28

Koeleria cristata 56, 66, 88
Koeleria macrantha 56, 57, 66, 67, 88, 89

lanatus, *Holcus* 52, 53, 62, 63
Lesser Hairy-brome 22
Leymus arenarius 18, 19
Lolium multiflorum 26, 27
Lolium perenne 68, 69
Loose Silky-bent 50
Lop Grass 64
Lyme Grass 18

macrantha, *Koeleria* 56, 57, 66, 67, 88, 89
marina, *Desmazeria* 48, 49, 82
marinum, *Catapodium* 48, 82, 83
marinum, *Hordeum* 26, 27
maritima, *Puccinellia* 80, 81, 94, 95
maritima, *Spartina* 14, 15
Marram Grass 28, 86
Marsh Foxtail 46
Mat-grass 84
maxima, *Glyceria* 70, 71
Meadow Barley 22
Meadow Brome 64
Meadow Fescue 24
Meadow Foxtail 44, 60
Meadow Oat-grass 76
Meadow-grass, Annual 78
Meadow-grass, Flattened 74
Meadow-grass, Narrow-leaved 88
Meadow-grass, Rough 78
Meadow-grass, Smooth 74
Meadow-grass, Spreading 74
Meadow-grass, Wood 78
media, *Briza* 60, 61
Melica uniflora 58, 59
Melick, Wood 58
membranacea, *Vulpia* 83
Milium effusum 40, 41
Millet, Wood 40
Molinia caerulea 14, 15
mollis, *Bromus* 64

mollis, Holcus 34, 35, 54, 55, 62, 63
montana, Agrostis canina 34, 38, 92
Moor-grass, Blue 72
Moor-grass, Purple 14
multiflorum, Lolium 26, 27
murinum, Hordeum 22, 23, 26, 27
myosuroides, Alopecurus 50, 51
myuros, Vulpia 83

Nardus stricta 84, 85
Narrow-leaved Meadow-grass 88
nemoralis, Poa 78, 79
notata, Glyceria 70, 71

Oat-grass, Downy 66
Oat-grass, False 30, 32, 50
Oat-grass, Meadow 76
Oat-grass, Yellow 40, 56
odoratum, Anthoxanthum 28, 29
ovina agg., Festuca 88, 89

Parapholis strigosa 48, 49
perenne, Lolium 68, 69
Perennial Rye-grass 68
Phalaris arundinacea 34, 35
Phleum arenarium 48, 49
Phleum pratense bertolonii 30, 31
Phleum pratense pratense 30, 31
Phragmites australis 12, 13
pinnatum, Brachypodium 36, 37, 42, 43, 54, 55
plicata, Glyceria 70
Plicate Sweet-grass 70
Poa angustifolia 88, 89
Poa annua 78, 79
Poa compressa 74
Poa humilis 74, 75
Poa nemoralis 78, 79
Poa pratensis 74, 75
Poa subcaerulea 74
Poa trivialis 78, 79
praecox, Aira 84, 85
pratense bertolonii, Phleum 30, 31
pratense pratense, Phleum 30, 31
pratensis, Alopecurus 44, 45, 60, 61
pratensis, Avenula 76, 77
pratensis, Festuca 24, 25

pratensis, Helictotrichon 76
pratensis, Schedonorus 24, 25
pratensis, Poa 74, 75
pubescens, Avenula 66, 67
pubescens, Helictotrichon 66
Puccinellia distans 80, 81, 92, 93
Puccinellia maritima 80, 81, 94, 95
pungens, Agropyron 18
Purple Moor-grass 14
Purple Small-reed 52
pycnanthus, Elymus 18

Quaking-grass 60

racemosus, Bromus 64, 65
ramosa, Bromopsis 20, 21
ramosa, Zerna 20
ramosus, Bromus 20
Red Fescue 86
Reed Canary-grass 34
Reed Sweet-grass 70
Reed, Common 12
Reflexed Saltmarsh-grass 80, 92
repens, Agropyron 18
repens, Elymus 18, 19
repens, Elytrigia 18
rigida, Desmazeria 48, 82
rigidum, Catapodium 48, 49, 82, 83
Rough Meadow-grass 78
rubra agg., Festuca 86, 87
Rye Brome 60
Rye-grass, Italian 26
Rye-grass, Perennial 68

Saltmarsh-grass, Common 80, 94
Saltmarsh-grass, Reflexed 80, 92
Sand Cat's-tail 48
Sand Couch 28
Schedonorus arundinaceus 24, 25
Schedonorus giganteus 24, 25
Schedonorus pratensis 24, 25
Sea Barley 26
Sea Couch 18
Sea Fern-grass 48, 82
secalinum, Hordeum 22, 23
secalinus, Bromus 60, 61

Sesleria albicans 72
Sesleria caerulea 72, 73
setacea, Agrostis 90
Sheep's-fescue 88
Sieglinglia decumbens 14
Silky-bent, Loose 50
Silver Hair-grass 84
Small Cord-grass 14
Small Sweet-grass 72
Small-reed, Purple 52
Small-reed, Wood 34
Smaller Cat's-tail 30
Smooth Brome 64
Smooth Meadow-grass 74
Soft Brome 64
Soft-grass, Creeping 34, 54, 62
Spartina anglica 16, 17
Spartina maritima 14, 15
Spartina x *townsendii* 16, 17
spica-venti, Apera 50, 51
Spreading Meadow-grass 74
sterilis, Anisantha 64, 65
sterilis, Bromus 64
stolonifera, Agrostis 46, 47
stricta, Nardus 84, 85
strigosa, Parapholis 48, 49
subcaerulea, Poa 74
Sweet Vernal-grass 28
Sweet-grass, Floating 72
Sweet-grass, Plicate 70
Sweet-grass, Reed 70
Sweet-grass, Small 72
sylvaticum, Brachypodium 58, 59

Tall Fescue 24
tenuis, Agrostis 36, 42
Timothy 30
x *townsendii, Spartina* 16, 17
Townsend's Cord-grass 16
Trisetum flavescens 40, 41, 56, 57
trivialis, Poa 78, 79
Tufted Hair-grass 38, 76

uniflora, Melica 58, 59
Upright Brome 62, 66

Velvet Bent 44, 94
Vernal-grass, Sweet 28
vinealis, Agrostis 34, 35, 38, 39, 92, 93
Vulpia 82, 83
Vulpia bromoides 83
Vulpia fasiculata 83
Vulpia membranacea 83
Vulpia myuros 83

Wall Barley 22, 26
Wavy Hair-grass 92
Whorl-grass 70
Wild-oat 56
Wood Barley 20
Wood Meadow-grass 78
Wood Melick 58
Wood Millet 40
Wood Small-reed 34

Yellow Oat-grass 40, 56
Yorkshire Fog 52, 62

Zerna ramosa 20